Geothermal Energy

Geotherm

Augusta Goldin

al Energy

A Hot Prospect

Illustrated with photographs, diagrams, and maps

Harcourt Brace Jovanovich, Publishers
San Diego   New York   London

The metric equivalents are approximate figures.

Copyright © 1981 by Kenneth Darwin Goldin

*Requests for permission to make copies of any part of the work should be mailed to
Permissions, Harcourt Brace Jovanovich, Publishers, Orlando, Florida 32887*

The quotation on page 70 from A. W. Reed's *Power from the Earth at Wairakei* is reprinted by permission of A. H. & A. W. Reed Ltd. and Longman
Paul Ltd. The poem on page 114 is by Oliver Houck, which appeared in
Ranger Rick's Nature Magazine, May 1974, and is reprinted by permission
of the National Wildlife Federation. The excerpt on pages 78–80 from the
paper "Study and Utilization of the Earth's Thermal Energy in the
U.S.S.R." by Dr. I. M. Dvorov of the Academy of Sciences, Moscow, is reprinted with the permission of the author.

LIBRARY OF CONGRESS
CATALOGING IN PUBLICATION DATA
Goldin, Augusta R
Geothermal energy.
Bibliography: p.
Includes index.
1. Geothermal engineering.   I. Title.
TJ280.7.G64   621.44   80-8800
ISBN 0-15-230662-5

C D E   Printed in the United States of America

To Dr. Joseph Barnea,

geothermal guru,

for enlightening me, directing me,

and curbing my enthusiasm when it

conflicted with economic feasibility

Contents

Geothermal Energy

1

Geothermal Energy Is

Would you believe that a mile or two below your feet there's more heat energy than the world could ever use? It's there in the form of steam, hot water, magma, and hot dry rocks. This is geothermal energy—"geo" meaning earth and "thermal" meaning heat. This earth heat is hidden inside the planet, but there are times when some of it comes to the surface and leaks out.

What good is this earth heat, this geothermal energy? For the ancients it was mainly a source of mystification. For today's geologists and engineers, however, it is a splendid alternative energy source.

Long ago people suspected there was natural heat inside the earth when they saw volcanoes blasting skyward, hot springs bubbling under falling snow, and geysers, smelling of brimstone, shooting into the air. Was it demons, they wondered, or angry gods, or the devil himself that stirred up this great heat? Magicians and sorcerers could offer no explanation. They could only dream and scheme and cast spells as they tried to conjure some of it to the surface.

Realistic Dream

Centuries later, science-fiction writers added to that dream. They talked of drilling holes deep down to the hot interior. They predicted that from those deep holes engineers would be able to extract the earth's natural heat and use it for the everyday heating and cooling of houses, chicken coops, schools, whatever; and for producing electricity to grind corn and coffee beans, saw lumber, light lamps, and spin music records.

Now this geothermal dream is coming true, and it just might help solve the energy crisis.

Presently, some fifty countries around the world are plugging into the earth, tapping its heat, and drawing some of it off in the form of steam and hot water. With this steam and hot water, some are running geothermal power plants and producing electricity. Others are building municipal thermal systems, which keep townspeople warm in winter and cool in summer. Still others are heating greenhouses and growing food and flowers for the market. Wherever geothermal energy is used, it turns out to be cheaper than oil or coal, natural gas or nuclear power. In San Francisco, California, and Larderello, Italy, the utility bills are very low. The same is true in Iceland, and in some parts of New Zealand, Mexico, and Japan.

It's no secret that there's a tremendous amount of heat locked up in the earth. If you could visit a mine or an oil well, you'd see it was so. Even in a shallow mine, as you walked to the lower levels, you'd begin feeling the way miners feel—pretty warm. If you visited a very deep mine (say a gold or diamond mine in Africa), you'd be surprised. First you'd hop into a shaft cage and drop down for a mile or two (1½ to 3 kilometers). Then you'd find that the mine down there had been air-conditioned. "Deep mines have to be air-conditioned," the uniformed guide would tell you, "because under normal conditions the earth's natural heat increases by as much as 50 to 100 degrees F (10 to 38 degrees C) with every mile of depth." He would also tell you that without air-conditioning, the temperature in some really deep mines might be a sizzling 200 degrees F (93 degrees C). It would be almost as hot down there as it is in a pot of water that's boiling on the kitchen stove.

You'd see another example of the earth's inner heat at an oil well. There you'd find that the oil welling up in the pipes is too hot to touch. At Bakersfield, California, for example, the oil that comes up from a depth of 9,000 feet (2,743 meters) is 241 degrees F (116 degrees C). Still, we have to remember that the earth heat flowing to the surface is only an indication of the superheat at the earth's center. There, according to scientific estimates, the temperature might reach 10,000 degrees F (5,537 degrees C).

The earth's superhot center is called the *core* and is about 4,400 miles (7,080 kilometers) in diameter. Scientists believe that this core is two-layered, the outer layer being composed of molten iron, the inner one of iron that has been solidified by pressure. The core is surrounded by a *mantle* of hot plastic or solid crystalline rock. Fortunately for the world, for the people and the other animals, and for all the plants there are, this hot mantle is insulated by a relatively cool *crust*.

The crust is the part of the planet we live on, work on, and play on. It is a thin layer that encircles the mantle to a depth of 5 to 40

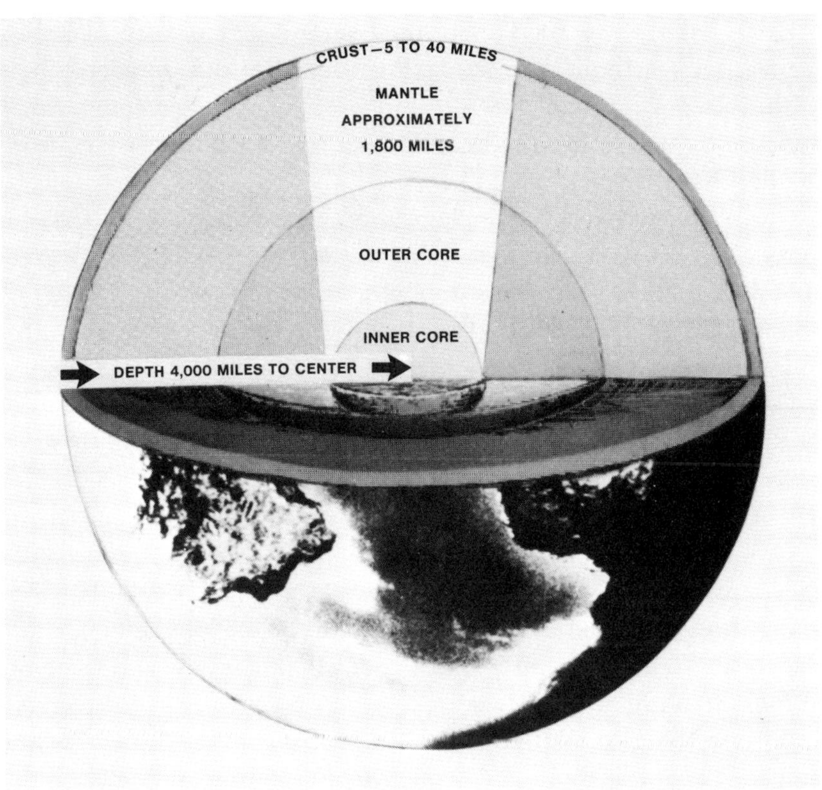

The earth's interior structure is concentrically layered. Shown here are the crust, the mantle, and the two-part core. (U.S. Geological Survey)

miles (8 to 64 kilometers). And while the mantle is made up of rock that hardly anyone has ever seen, the crust (with its mountains and valleys, its beaches and oceans and deserts and icecaps) is familiar to all of us. This crust is made of rocks and sand, gemstones and metals, water and other minerals, as well as gases and hot magma usually found beneath the surface. (Magma is volcanic rock in its molten stage.)

If you could slice through the earth the way you slice through an onion with its concentrically layered rings, you would see that it looks something like the diagram above.

Earth scientists learned about the contents of the earth by studying seismic (earthquake) waves, which move at different velocities as they travel through different kinds of rocks, liquids, and gases, and also by measuring the heat flowing to the earth's surface from the depths.

One of the first great seismologists, a Yugoslav named Andrija Mohorovičić, observed a strange phenomenon concerning these waves. He found that they showed a dramatic change in velocity at a depth of about 22 miles (35 kilometers) below the continents and about 6 miles (10 kilometers) below sea level. He called this a discontinuity, a dividing line between the lowest level of the crust and the mantle. This discontinuity was subsequently named the Moho in his honor.

Scientists have long wanted to pierce the Moho with deep holes and are often heard to grumble because funds are inadequate. If our government can support the exploration of space for hundreds of thousands of miles, they say, surely it can support our drilling a Mo-hole (a hole through the Moho) 3 to 5 miles (5 to 8 kilometers) below the ocean floor so we can find out about the heat flow and the kind of rocks that are down there.

It's one thing to know there's all that heat inside the earth, but how that heat got there is still unclear. For centuries, philosophers and scientists assumed this heat was residual—retained from the time the earth was formed. More recently, geologists have been offering other theories and explanations. The two most frequently accepted today are:

- that the core gets its great heat from pressure
- that radioactive elements, which are present in some rocks, are constantly disintegrating and releasing tiny amounts of heat.

Uranium and thorium (from which nuclear fuels are refined for atomic bombs) decay and turn into lead; potassium decays and turns into argon; carbon decays and turns into nitrogen; rubidium decays and turns into strontium, the stuff that poisoned our milk some years ago when atomic test bombs were exploded in the Pacific.

This decay of radioactive elements in the rocks has been going on for billions of years, and the released bits of heat have been adding up. No wonder some people see the interior of the earth as an ongoing radioactive furnace. Some go so far as to say that this makes our earth a planetary nuclear reactor.

Does this mean you are actually standing on a planetary nuclear reactor?

Yes, it does, but do not worry. Although radioactivity is constantly going on, it's low level, the kind you can live with. Furthermore, we are generally shielded by the cool, hard crust we live on—except in places where radioactive mineral concentrations are great and the crust happens to be weak. There earth heat leaks through. Such places are called *hot spots.*

Hot spots pepper the globe, and wherever they occur, no one has to be told, "Hey, there's a hot spot!" because it's easy to hear the volcanoes rumble and to see the gases billow, the hot springs bubble, the little geysers splash, and the big ones erupt with great violence. Still, even the most violent hot spots are rarely dangerously radioactive.

There's yet another source of heat that adds to the earth's thermal inventory. This source is friction. Now friction is, actually, part of our everyday life. When we rub our hands together, they get warm. When we saw a piece of wood, the saw blade gets hot. When we grind the edge of a knife in an electric sharpener, the sparks fly. When we race a car for several miles, the tires heat up. All these are examples of friction.

In the earth, a certain amount of heat is generated by friction, as well as by other sources, at the boundaries of the *crustal plates.* These plates are huge sections of the earth's crust that, in places, extend into the mantle. There are a number of major plates and a lesser number of smaller ones, as you can see on the map on pages 18–19. These plates carry the continents and oceans in constant slow motion—at the rate of about 2 inches (5 centimeters) a year. Sometimes these plates bump into each other at the boundaries. Sometimes they pull apart. Other times they override each other like ice slabs in a frozen river.

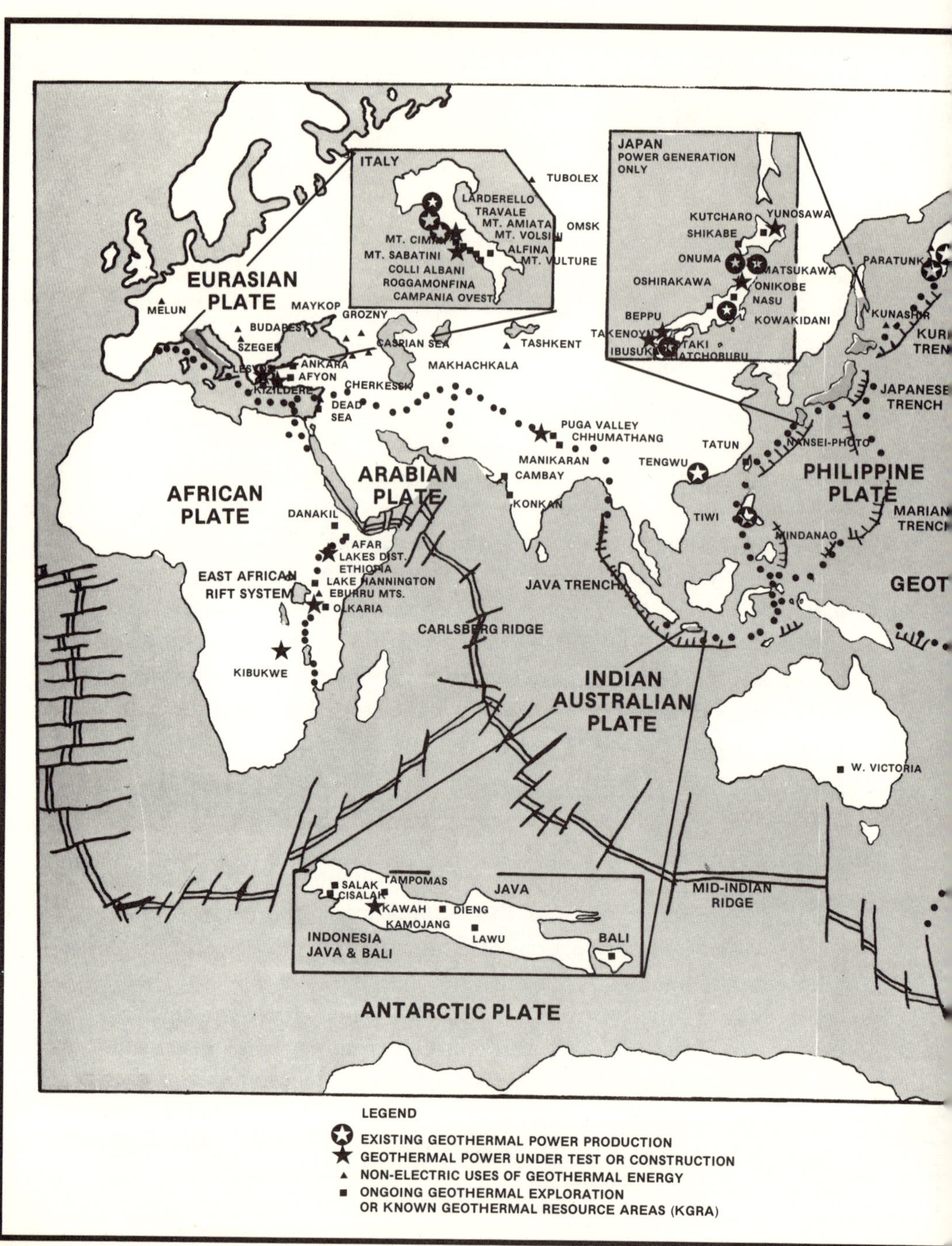
ITALY
LARDERELLO
TRAVALE
MT. AMIATA
MT. VOLSINI
MT. CIMINI
MT. SABATINI
MT. VULTURE
ALFINA
COLLI ALBANI
ROGGAMONFINA
CAMPANIA OVEST
TUBOLEX
OMSK
JAPAN
POWER GENERATION
ONLY
KUTCHARO
YUNOSAWA
SHIKABE
ONUMA
MATSUKAWA
OSHIRAKAWA
ONIKOBE
NASU
BEPPU
KOWAKIDANI
TAKENOYU
TAKI
IBUSUKI
HATCHOBURU
EURASIAN
PLATE
MELUN
MAYKOP
GROZNY
BUDAPEST
SZEGED
LESVOS
ANKARA
AFYON
KIZILDERE
CHERKESSK
CASPIAN SEA
TASHKENT
MAKHACHKALA
DEAD
SEA
AFRICAN
PLATE
DANAKIL
ARABIAN
PLATE
PUGA VALLEY
CHHUMATHANG
MANIKARAN
CAMBAY
KONKAN
TATUN
TENGWU
NANSEI-PHOTO
PHILIPPINE
PLATE
PARATUNK
KUNASHIR
KURIL
TRENCH
JAPANESE
TRENCH
MARIANA
TRENCH
TIWI
AFAR
LAKES DIST.
ETHIOPIA
LAKE HANNINGTON
EBURRU MTS.
OLKARIA
EAST AFRICAN
RIFT SYSTEM
KIBUKWE
JAVA TRENCH
CARLSBERG RIDGE
INDIAN
AUSTRALIAN
PLATE
MINDANAO
GEOT
W. VICTORIA
MID-INDIAN
RIDGE
TAMPOMAS
SALAK
CISALAH
JAVA
KAWAH
DIENG
KAMOJANG
LAWU
BALI
INDONESIA
JAVA & BALI
ANTARCTIC PLATE
LEGEND
EXISTING GEOTHERMAL POWER PRODUCTION
GEOTHERMAL POWER UNDER TEST OR CONSTRUCTION
NON-ELECTRIC USES OF GEOTHERMAL ENERGY
ONGOING GEOTHERMAL EXPLORATION
OR KNOWN GEOTHERMAL RESOURCE AREAS (KGRA)

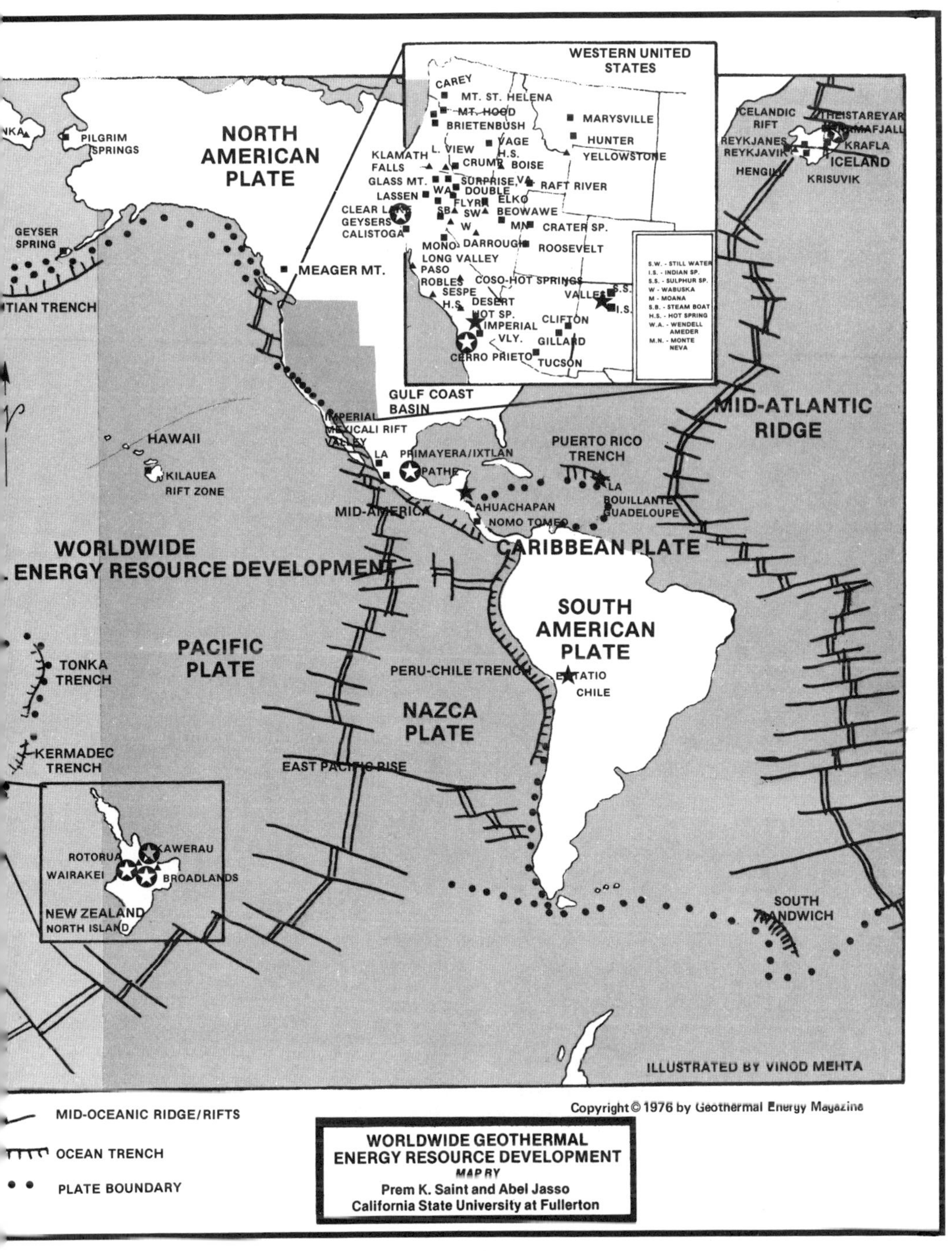

NORTH AMERICAN PLATE
PILGRIM SPRINGS
NKA
GEYSER SPRING
TIAN TRENCH
HAWAII
KILAUEA RIFT ZONE
WORLDWIDE ENERGY RESOURCE DEVELOPMENT
PACIFIC PLATE
TONKA TRENCH
KERMADEC TRENCH
EAST PACIFIC RISE
ROTORUA
KAWERAU
WAIRAKEI
BROADLANDS
NEW ZEALAND NORTH ISLAND
WESTERN UNITED STATES
CAREY
MT. ST. HELENA
MT. HOOD
BRIETENBUSH
MARYSVILLE
HUNTER
YELLOWSTONE
VAGE H.S.
L. VIEW
KLAMATH FALLS
CRUMP
BOISE
GLASS MT.
SURPRISE V.
RAFT RIVER
LASSEN
DOUBLE
WA
FLYR
ELKO
SB
SW
BEOWAWE
CLEAR L.
GEYSERS
CALISTOGA
W
MN
CRATER SP.
MONO
DARROUGH
ROOSEVELT
LONG VALLEY
PASO ROBLES
SESPE
COSO-HOT SPRINGS
VALLES
S.S.
I.S.
MEAGER MT.
DESERT HOT SP.
H.S.
CLIFTON
IMPERIAL VLY.
GILLARD
CERRO PRIETO
TUCSON
S.W. - STILL WATER
I.S. - INDIAN SP.
S.S. - SULPHUR SP.
W - WABUSKA
M - MOANA
S.B. - STEAM BOAT
H.S. - HOT SPRING
W.A. - WENDELL AMEDER
M.N. - MONTE NEVA
GULF COAST BASIN
IMPERIAL MEXICALI RIFT VALLEY
LA PRIMAYERA/IXTLAN
PATHE
MID-AMERICA
AHUACHAPAN
NOMO TOMEO
PUERTO RICO TRENCH
LA BOUILLANTE
GUADELOUPE
CARIBBEAN PLATE
MID-ATLANTIC RIDGE
ICELANDIC RIFT
REYKJANES
REYKJAVIK
HENGILL
KRISUVIK
THEISTAREYAR
NAMAFJALL
KRAFLA
ICELAND
SOUTH AMERICAN PLATE
PERU-CHILE TRENCH
TATIO CHILE
NAZCA PLATE
SOUTH SANDWICH
ILLUSTRATED BY VINOD MEHTA
MID-OCEANIC RIDGE/RIFTS
OCEAN TRENCH
PLATE BOUNDARY
Copyright © 1976 by Geothermal Energy Magazine
WORLDWIDE GEOTHERMAL ENERGY RESOURCE DEVELOPMENT
MAP BY
Prem K. Saint and Abel Jasso
California State University at Fullerton

Ready to drill down a mile or more for steam or hot water. (U.S. Department of Energy, photo by Jack Schneider)

It's at the crustal boundaries that the crust is thinnest. There you will find plenty of geothermal energy leaking through. There you will find hot spots such as those in the Pacific Ring of Fire, which contains 80 percent of the world's active volcanoes, hot springs, geysers, and fumaroles, as well as hot-water fields.

To trace the Ring of Fire, turn to the map (pages 18–19) and put your finger on the southwest tip of South America, move it northward to the coast of Central America, to Mexico, to North America, and all the way to Alaska. At that point, continue westward to Siberia, then southward to the Japanese Archipelago, and down to Southeast Asia. This Ring of Fire is peppered with geothermal fields.

Until the twentieth century, the only clues to the presence of the earth's interior heat were the visible heat leakages. These leakages were plentiful, but their significance was not known, because nobody had any idea that geothermal energy could be controlled. Nobody knew that deep holes by the mile could be drilled into the interior, that these holes might fill up and become hot-water wells or steam wells, and that these wells could be developed to power geothermal electric generating stations or to heat and cool buildings.

Now we come to the end of the twentieth century when geologists, chemists, and engineers know a great deal about locating geothermal reservoirs and about drilling into these reservoirs. They dream now of geothermal development on a large scale in order to provide abundant and inexpensive energy for man's everyday use.

They dream of great cities that will draw geothermal energy from hot-water wells and steam wells. They dream of tiny isolated villages in the frozen Arctic and the torrid jungles that will be electrified for the first time in history. They dream of underground cities powered geothermally. They dream of geothermal energy for the people—energy that will last a long time and is clean, abundant, and more economical than the conventional forms of energy.

And with modern technology and expertise, they know that the dream of developing geothermal energy is a realistic one.

2

Some Like It Hot!

Today people are inclined to think big. If geothermal hot water is available, nothing less than big heating and air-conditioning units are considered by investors. If geothermal steam is available, then it's a steam-shrouded power station with its snaking steam pipes. If, however, you could go back in time, you would find people thinking small and glad to employ the hot water for personal uses wherever they found it coming freely out of the ground. Steam, however, was a no-no, a frightening monster.

In the long ago, when roaming hunters and wandering shepherds chanced across a hot spring or a geyser, they fled. The more venturesome stopped and sniffed. Sometimes, they even poked a finger into the water, but then they too fled, convinced it just wasn't natural for water that came out of the ground to be hot.

Natural or not, this hot water tantalized children. Where it wasn't actually scalding, spunky boys and girls would splash and dangle their feet in it. They'd jump into the pools and laugh as the mud rinsed off their bodies.

In time, women began to look at the springs with cautious interest. They learned that eggs, tied in a couple of leaves and placed in
a scalding spring, became hard cooked. They learned that a chick-
en, along with barley, onions, carrots, and herbs, packed in a water-
tight grass basket and lowered next to the eggs, stewed tender and
delicious. Little girls discovered that the berries picked in the fields
made great puddings. A mixture of juicy berries, a bit of wild honey,
a sprinkle of rice flour, plus a few shredded sweet-smelling leaves,
when crammed into a gourd, thickened into a tasty treat next to the
chicken in the same hot spring.

In the areas now called Asia, Africa, the Americas, and Australia,
people gradually learned to use the hot water that came naturally
out of the ground to make life easier for themselves. The men began
heating their homes with this hot water, sometimes building huts
and bath houses right over the springs. In some places two or three
families, using pointy sticks, would dig little channels to carry the
hot water from the springs right through their earthen floors.

Then sometimes personal miracles seemed to occur. People discovered that a good soak in those hot springs (cooled if necessary in overflow tubs or ponds) cured certain ailments. Stiff muscles and sore backs became limber. Skin diseases cleared up. Wounds healed.

According to American archaeologists, the Indians knew about the curative powers of geothermal hot water in the area now called Hot Springs, Arkansas. As long ago as 6,000 B.C. (and some push the date back to 12,000 B.C.), the Indians considered those waters sacred—so sacred that embattled warriors, belonging to different tribes and speaking different tongues, used to gather there, lay aside their arms, and bathe in peace. In 1541, when Hernando De Soto discovered these waters, he found the legends still alive in Indian folklore. Later, European immigrants by the thousands traveled there on horseback and in covered wagons to sample the curative powers of those hot springs. One, which they named Corn Hole Spring, was particularly popular. Here, in their wrinkled clothes, visitors with foot problems would sit on wooden benches and try to soak away their aching corns in the hot water.

In the 1880s, the United States government recognized the therapeutic value of those hot waters. It established and operated the Army and Navy General Hospital there for servicemen and women (and even for some prisoners of war) who needed treatment for arthritis, rheumatism, and polio. This facility was closed down some eighty years later, but thousands of patients continue to come to private hospitals: St. Joseph's, Quachita Memorial, and Leo N. Levi Memorial.

Through the years chemists kept talking about the hot spring waters as being beneficially radioactive and, indeed, medicinal, and testimonials concerning their value came from around the United States. A story about Oregon's hot springs in the *Ashland Tidings* of August 10, 1876, said: "Mr. Brook has erected a comfortable bath house supplied with water from one of those springs. In order to reduce the temperature within limits of human endurance, he has erected a large tank which he fills with water, and lets it stand until it cools. . . . By drinking and bathing in these waters, great relief and, in some instances, perfect cures have been effective to those troubled with kidney diseases."

A more recent account that appeared in the *Oregon Journal* on January 8, 1939, described an interview with an old Indian woman who told about the ancient practice of carrying the aged on litters over the mountains so they could be cured by the cooler overflow of Oregon's Big Spring.

Relief and cures were also reported some 2,500 years ago by the ancient Greeks. At that time, those who could afford it used to sail to the island of Lesbos for literary and philosophical conferences. Here, while participating in all that good talk, they enjoyed the hot healing waters in the steamy stone huts with the rounded roofs. See the photograph at the top of page 26.

And far to the north in Bath, England, there were also geothermal springs reputed to be sacred and medicinal. When the Roman legions swept over the countryside about 2,000 years ago, they immediately erected bath houses for therapeutic bathing. In the hot-water pools, the legionnaires would soak their aching muscles and stiff joints. In the cold-water plunge pools, they would invigorate their spirits. So popular was geothermal bathing that wherever their conquering armies marched, the Roman engineers had only to see vapor curling away from a hot spring and—presto—there was a bath house. Some few still remain, in total disrepair, along the northern shores of Africa. But one, near Tripoli, was excavated not long ago in a fairly good state of preservation after having been buried under desert sands for something like 1,500 years. See the photograph at the bottom of page 26.

Wherever there were hot springs, people found uses for the water. In New Zealand, the Maoris, from the day they grounded their canoes after their historic voyage from Polynesia, used hot springs for cooking and heating and pleasurable bathing, as well as for curing what ailed them. Even today New Zealanders and Azoreans enjoy a meal cooked in a hot spring, as do also the Japanese.

Hot-spring bathing has always been considered a traditional plea-

Above: Ancient Grecian geothermal bathhouses still in use on the island of Lesbos. (From *Basics of Applied Geothermal Engineering*, Edward F. Wehlage, P. E.)   *Below*: Ancient Roman geothermal bathhouses on Lepcis Magna, near Tripoli, which had been buried under the desert sands for nearly 1,500 years. (From *Basics of Applied Geothermal Engineering*, Edward F. Wehlage, P. E.)

Maoris cooking in a hot pool in Rotorua, New Zealand. (NPS)

sure and a highly religious experience in Japan. In a letter to this author, Tetsumaro Senge, chairman of the National Parks Association, touches on the legends surrounding these springs. "There are many Deer Springs, Bear Springs, Crane Springs, etc.," he writes, "named after animals or birds because, long ago, hunters and other people found these animals in the springs curing their wounds. People were taught that the springs were very useful for keeping health and curing wounds, and they worshiped the power of the springs and bathed with many thanks."

As long as the Japanese can remember, their sick and ailing have journeyed to the *kamiyu* (divine bath). But the healthy also came: for a good visit with their friends, a jolly feast, a healthful dip, and a chance to sleep on the beach in a *sunayu*. A sunayu is a sand hot spring where the steam curls up through the sand. There the vis-

Azoreans in São Miguel using a hole-in-the-ground oven. In three hours the hot geothermal vapors cook pork, kale, and cabbage inside metal containers in the bags. The metal containers keep the sulfur smell out of the food. (© National Geographic Society, photo by O. Louis Mazzatenta)

itors dig themselves in, leaving only their heads exposed, and so sleep away the night.

Most of Japan's hot springs are enclosed within the boundaries of national parks, where many attractive spas have been developed. Of these parks, none is more famous than Nikko, which was founded as a sacred geothermal sanctuary. "Nikko National Park," writes Senge, "is called the place where both natural and artificial beauty are tightly combined. There are many mountains, lakes, waterfalls, forests, and plains with big shrines and temples decorated with beautiful paintings and fine carvings. Some 1,200 years ago the Bud-

Japanese bathing in a riverside hot spring. (Kiyoshi Takahashi)

dhist saint, Shodo Shonin, climbed to the top of Mount Nantai and worshiped there. He also found a hot spring in the Nikko Mountain area, which refreshed the spirits and restored the bodies of his religious people."

Today, according to Senge, "Hotels, inns, and lodges concentrate around hot springs, and most have a big bath that can accommodate fifty to two hundred people."

Each of these big baths, as well as the smaller bathing sites, besides offering pleasure to the visitors, holds out a promise of different cures. Varying in temperature and in chemical composition, these hot-spring waters also vary in their reputed curative powers.

There are carbonic acid springs said to be good for stomach troubles; sulfur springs for improving the blood's circulation; alkali and bicarbonate of soda springs good for burns and wounds and skin diseases; sulfur and alum springs also for skin problems; and radium

A big, modern bath at the Hotel New Okabe in Nikko National Park. Most Japanese hotels and inns have two big baths: one for men and one for women. (Mr. Norio Saito)

    Geothermal Energy

springs said to relieve (and in some instances to cure permanently) certain forms of arthritis. There are also strong acid springs that inflame the skin and are believed to cure various kinds of venereal diseases. In all, there seem to be hot springs for every purse and every ailment.

On Japan's southern island, the springs at Beppu offer relief for an amazing assortment of ailments to some 12 million visitors a year. There every day some 27,000 gallons of mineral water, varying in temperature from fiery hot to medium warm, flow out of more than 3,000 openings. After people have sampled the hot baths, they are usually told to "go to a hell." A Beppu hell is a noisy jigoku—a boiling pond that detonates loudly every once in a while. There are quite a few jigoku there, all sourrounded by muds of many colors—blue, green, slate, vermilion, whatever—all very inviting to the children who love to play in the mud.

During the nineteenth century many hydros—watering places—began to appear along the Mediterranean coast. Crowds patronized these hydros because, aside from their healing properties, they were fashionable. They provided the best of holidays for people, who came with their luggage and their servants "to take the waters" for a few weeks every year.

After World War I, "taking the waters" in Europe declined in importance for two reasons. Low-cost oil and gas made it possible for people to have plenty of hot water in their own homes. Although this water doesn't provide the therapy that chemically charged geothermal waters do, it was ever so convenient to be able to soak in a hot tub at home. By this time, the pharmaceutical companies were able to provide instant pills for instant relief.

Around the world, however, the poor continued to use the hot-spring waters in the same old way. In El Salvador and Mexico, village women still wash their clothes in warm streams. In India, some still channel hot-spring water along the floor of their huts because this means the difference between a cold, dank hut and one that's toasty warm. In the Azores, people unable to afford doctors still treat their ailments with hot-spring water and mud packs.

Women of Puerto Vallarta, Mexico, washing their clothes in the Cuale River. (United Nations/H. Bijur)

Currently, a number of facilities around the world provide geothermal water treatments. Among them, and most famous, are the three previously mentioned prestigious hospitals at Hot Springs, Arkansas. Another famous one is Queen Elizabeth Hospital located in Rotorua, New Zealand.

In a letter to this author, Dr. I. C. Isdale, the medical superintendent there, writes, "At the Queen Elizabeth Hospital we have maintained the use of hydrotherapy [water treatment] for various rheumatic complaints. We still use both the alkaline 'Rachel' waters and the acid 'Priest' waters in immersion pools, and we use the local silicate mud for mud baths and mud packs as an effective way of applying heat deeply." He then adds, "We have not conducted scientific tests of their efficacy but have an established place for the use of hydrotherapy for the many patients who come to Rotorua with chronic rheumatic problems."

With all those millions of barrels of hot water pouring out of the earth's crust every day, one can't help wondering why some grand geothermal projects, other than spas, were not developed prior to World War I.

Well, one project was developed. It was big, it was successful, and it was astonishing because, despite its size, no machines were required.

The year was 1901, the time of the Alaska gold rush. The place, a drear, uninhabited snowfield, 120 miles south of the Arctic Circle.

It was the gold rush that brought prospector J. F. Karshner up there. He discovered not gold but something better than gold—a few small hot springs bubbling out of a depression between two hills. Karshner staked out the land, built a cabin, and moved in. Soon an interested party appeared—Frank Manley, a Texan. Manley had struck it rich at the Cleary Creek diggings, and now, with a couple of hundred thousand dollars' worth of gold nuggets bulging his pockets, he looked over the Karshner homestead. Seeing that it was within hollering distance of a newly slapped-together tent city of prospectors, he proposed a business deal. In that frozen land the two men, Karshner and Manley, created the Manley Hot Spring Plantation Resort—a sixty-room hotel with hot running water piped in from the springs. They enclosed a number of bathing pools of varying temperatures for the grimy prospectors. They constructed a dairy, poultry coops, and hog barns, and raised feed grain on the hillsides. They managed garden plots where the ground was warmed by nearby springs, and built geothermally heated greenhouses for growing vegetables the year round. And they made money. In fact, in 1910, they realized, so it was said, something like $300,000 just from their potato crop of 150 tons, which they sold at a dollar a pound to a snowbound mining camp nearby.

Successful gold-rush prospectors visiting the corn fields at Manley Springs. Note the built-up stone curbing around the hot spring (*lower left*). (Archie Lewis Collection/VF in the Archives, University of Alaska, Fairbanks)

Then the gold rush petered out. World War I came along, the Manley Hotel burned to the ground, and the Manley geothermal experiment was no more.

Why had that experiment succeeded so extravagantly? It succeeded because the hot springs were on high ground, and the hot water that served the hotel, dairy, coops, barns, and greenhouses came through the pipes by gravity flow. No pumps were used for this commercial operation, and very little technological expertise was required.

For many of today's geothermal operations, pumps and other specialized machines are definitely required. For this reason, geothermal development had to wait until the technology was good enough and cheap enough to deliver the hot water in any direction from the springs: uphill, downhill, or sideways. Equally important is the fact that geothermal projects on a commercial scale had to wait until the technology made it possible to locate hot-water reservoirs, to drill wells, and to bring up high-temperature water and steam from the interior of the earth. Projects also had to wait until geothermal energy was competitive in cost with fossil fuel energy.

We now come to the next stage in the geothermal story—to the utilization of the earth's steam and hot water for twentieth-century living—which begins with the harnessing of geothermal steam for the production of low-cost electricity.

3

Harnessing Dry Steam

Flip the switch in your kitchen and the lights go on, but where the electricity comes from is another story. In New York it may come from Niagara's hydroelectric station, or it may come from a nuclear facility. In Illinois, it may come from a generating plant fired by coal. In Texas, it would likely be from one fired by oil. In California, it may be geothermally powered.

The commercial development of geothermal electricity is a twentieth-century phenomenon, with an increasing number of power plants coming on stream in countries around the world. Some nations run those plants with the earth's natural steam, some with the earth's hot water. In Italy, the United States, and Japan, geothermal power plants send electricity through high-voltage lines to town and country. In Indonesia, El Salvador, and the Philippines, geothermal electricity serves mainly areas that had never seen electrical outlets before.

Only since 1904 has the earth's natural steam been harnessed and converted to electricity, and this is how it came about. It started in

for Electricity

the steam fields of northern Italy, in Larderello, where a number of boric-acid factories had been doing a thriving business. The manager of these factories was Prince Piero Ginori Conti, a most excellent businessman. It occurred to him one day that this steam, which was such a valuable source of dissolved boric acid, as well as ammonium carbonate and sodium carbonate, might be even more valuable as a source of power, because blowing steam has energy. He thought that if he could get it into a pipe and pipe it to one of those newly invented steam engines called a turbine, the steam would spin the turbine, the turbine would turn a generator, and the generator would produce electricity.

Conti corralled some clever engineers, some skillful geologists, and his money-minded board of directors. They talked about this steam that was blowing out of the fields. They agreed there was plenty of steam, but the question was how to handle it, control it, and direct it to the turbine.

It was decided, finally, that a steam well should be drilled to deep-

Vapor and gas blowing out of a steam field in the Larderello geothermal region. (By permission of the Italian National Electricity Board—ENEL)

en the steam hole and make sure of a steady supply. The drilling rigs moved in, the casings went down, and when the hole penetrated the geothermal reservoir, a burst of steam—mixed with bits of rock and sand and gravel—roared out of the borehole with an unearthly clatter. Then the workers with their hard hats and insulated gloves moved in. They constructed a wellhead with a turbogenerator unit. The pipes caught the steam, and the little power plant the prince had prepared worked just as he had imagined it—the steam spun the turbine, the turbine turned the generator, and the generator produced electricity. It actually lit up five electric light bulbs.

"Big deal," you say. No big deal, but a big breakthrough, for now scientists from around the world flocked to Larderello, asking, "How did you do it? How was that steam well drilled? How was the steam piped?" One thing they soon learned was that the prince had been lucky. The kind of steam he piped from his well was a rare kind called dry steam. What makes dry steam so desirable is that it contains very little water and is relatively clean.

Soon, however, the visitors learned that the Larderello dry steam didn't exactly rate A No. 1 because it wasn't clean *enough*. It wasn't dirty with dirt, mind you. This excellent steam was dirty with dissolved chemicals. They found this out when the five light bulbs flickered, dimmed, and went black as the turbine slowed down and wheezed to a stop.

The turbine stopped because the wheels couldn't turn. Their blades had become corroded and heavily scaled by the chemicals that were dissolved in the steam. It seemed that some of those chemicals, which were fine for the chemical business, were murder to the metal turbine blades.

The prince realized he had two options if he wanted to keep his little geothermal power plant going: he would either have to stockpile spare turbine wheels and blades so he could replace the damaged ones every so often; or he would have to clean up the dirty steam—remove the corroding and scaling chemicals and get rid of the few bits of sand and rock gravel before the steam entered the turbine.

Technical improvements went on for years. It wasn't until 1913, almost ten years after start-up, that the engineers were satisfied that the steam was clean enough. In a new, improved power plant, the switch was thrown, the steam whooshed through the pipes, the turbines turned, and this time the plant produced 250 kilowatts (250,000 watts) of electricity.

But what's a 250-kilowatt plant even if it is the only one of its kind in the world? It would illuminate only a thousand 250-watt light bulbs or run 200 portable heaters. By 1930, after more improvements, the little Larderello plant was generating a respectable 139 megawatts of geothermal electricity. (A megawatt is equal to 1,000

kilowatts and would serve 1,000 people at American standards.)
Then came World War II and trouble. The power plant was bombed
and demolished. Of course it was subsequently rebuilt and is now
bigger and better. Today, there's a complex of power stations at
Larderello. They have an installed capacity of more than 400 megawatts. They supply electricity for the Italian power grid, and they
run the electric railroads. Additionally, other smaller geothermal
power plants continue to boost that country's electrical output.

Italy is not the only place in the world where dry steam is used
to generate electricity. By the 1980s, there were three. The most
famous continued to be the one at Larderello. The second and larg-
est is in the United States—in California. The third and most recently developed is in Japan.

Believe it or not, the American steam fields, which are called The
Geysers, are misnamed. There are no geysers in those fields, only
hot springs and fumaroles, which are holes in the ground that emit
steam steadily. Geysers are altogether different: they send up foun-
tain-like jets of steam and hot water intermittently, as does Old
Faithful in Yellowstone National Park.

For centuries, only the Indians knew about The Geysers area,
which they called the Beautiful Land. Many settled there so they
could bathe in the hot springs and steam-cook their food in the fumaroles. They also used the mud that surrounded the springs. This
mud had special healing properties, but it was so hot that people,
crossing the area, had to curl up their toes and hop.

The Geysers were first seen by a white man in 1847—two years
before the California gold rush. William Bell Elliott, explorer by
temperament and surveyor by profession, was out one day tracking
bear. In the mountains between Cloverdale and Calistoga, he
topped a ridge and froze. There, before him, puffs of steam were
jetting out of a hillside. Plumes of steam were shooting skyward. Everything smelled of sulfur and brimstone, and Elliott, convinced he
had reached the gates of hell, was ready to say good-bye to the
world.

Word about this weird place spread quickly, and just as quickly,

Right: Men operating the rotary table of a drilling rig at Larderello. (By permission of the Italian National Electricity Board—ENEL)

Below: Geothermal power plant and cooling towers in the Larderello region. (By permission of the Italian National Electricity Board—ENEL)

Bird's-eye view of The Geysers, circa 1880.
(Courtesy of the California Division of Mines and Geology)

ATHS

promoters rushed to establish spas, American-style. They built cozy cottages and fancy hotels around the hot springs at Calistoga and at The Geysers, just a few miles away. They advertised mineral water for drinking and for medicinal bathing, in pools fed variously by sulfur springs and alum springs, and springs containing sulfate, magnesia, silicic acid, and Epsom salts. They arranged for horseback trips to the spas, and later, in the 1880s, for comfortable stagecoaches each drawn by a "six in hand." Crowds swarmed over the twisting, clay-packed roads to take the waters, to hop and frisk over the "oven grounds" (the Indian name for the hot muds), and to cure themselves of their disorders. Then, just after the turn of the twentieth century, the busy spas declined in importance. Following their European counterparts, American pharmacies began offering quick pills for quick relief. American engineers began poking into the ground, searching for geothermal reservoirs, and the spotlight swung to the production of geothermal electricity.

By 1922, drillers and drilling rigs were moving into The Geysers to tap the steam source. They installed steam pipes, pre-turbine "reciprocating engines," and generators, and produced electricity. Jubilation, celebration parties, flags, excitement, and newspaper headlines! But the venture was short-lived because of some of the same problems that beset the early Larderello venture: the pipes and turbines could not stand up to the corrosive and abrasive effects of the steam, of the pebbles, and the rock fragments that roared out of the steam well. One day the power plant would be working; the next it would be wheezing and groaning because the metal parts had been eaten up chemically, been pitted, and been covered with scale. And stainless steel alloys had not yet been invented. Furthermore, there were plenty of hydroelectric sites still available that were cheaper to develop than geothermal sites, so The Geysers project was abandoned.

By 1955, however, the word at The Geysers was "Go." With good steel alloys suddenly available, the business of producing geothermal electricity was on its way. Now busloads of geologists, engineers, and drillers, followed by the heavy equipment—load trailers and wheel loaders, generators, steel tanks, crawlers, and pipes—began rumbling up the same crooked roads that had served the spa patrons a hundred years earlier. Now two pioneering companies, Magma Power of Los Angeles and Thermal Power of San Francisco, began drilling for steam, and the Pacific Gas and Electric Company bought the steam and began generating electricity in its brand-new 11-megawatt power plant. Since 1967, they've been drilling, on the average, more than one new well per month, and these wells sometimes reach down to a depth of 1 to $1\frac{1}{2}$ miles (about $1\frac{1}{2}$ to $2\frac{1}{2}$ kilometers).

Still, problems continued to surface. The new alloys were not completely resistant to the damaging rock fragments in the rushing steam. These fragments punched tiny holes in the turbines, but ingenious engineers soon took care of that. They passed the steam through centrifugal separators, where the tiny bits and pieces were whirled off and out. Then the cleaned steam continued on its way through the pipes to the power plant, where electricity was generated as planned.

Mentally, it's easy to follow the generated electricity to the transmission lines, but this isn't the end of the story because other problems were still unsolved. Heading the list was the matter of the used or spent steam that is exhausted from the turbines. What's to be done with it? Release it to the atmosphere? The Environmental Protection Agency (EPA) would have something to say about that. It would immediately bring suit for damage to the environment because that spent steam is still very hot and it smells. It smells like rotten eggs because of the dissolved hydrogen sulfide (H_2S), a very poisonous gas that is contained in it. After much brainstorming around conference tables and trying out all sorts of ideas in the field, engineers hit on a workable solution. They decided to pipe that spent steam to a condenser. There it was liquefied and turned to water. Of course, the resulting water was hot, so it was piped to a cooling tower, where it was further cooled by evaporation. After that, some of the now cold water was utilized to chill the condenser for its continued use.

But what were they to do with the surplus water that was pouring from the cooling tower? There was so much that it threatened to swamp the area and pollute the nearby creeks. Well, they injected that surplus water into the ground and in this way replenished some of the underground supply, lengthened the life of the reservoir, prevented the ground from sinking, and safeguarded the environment.

This looked like a neat solution. It took care of the cooling steam, and it took care of the surplus water. But that pesky H_2S gas was still another problem. It was still poisonous, and it still smelled.

A later solution attacked this problem head on with "scrubbers" that were designed to clean the steam and eliminate up to 90 per-

Unit 11, which is one of the fourteen power plants at The Geysers, has a capacity of 106 megawatts. The small white-topped building in the right foreground contains the equipment for the reduction of hydrogen sulfide emissions. (Pacific Gas and Electric Company)

Called the Valley of Steam, this part of The Geysers geothermal steam field generates 800 megawatts of electricity from underground steam. It is 90 miles northeast of San Francisco. (Pacific Gas and Electric Company)

cent or more of the hydrogen sulfide. A scrubber is a device that's attached to the exhaust side of the turbine. There it separated the H_2S into its components: hydrogen and sulfur. Hydrogen, being a harmless gas, was allowed to escape into the atmosphere. The sulfur, having considerable commercial value, was extracted in pure form and sold to the chemical industries.

Still those scrubbers were not and are not 100 percent perfect. Enough of that rotten-egg smell still lingers around the plants to cause the government inspectors to keep a sharp eye on emission standards. As Michael J. Miller, air specialist who had served at the Pacific Gas and Electric as administrative assistant, wrote to this author, "To meet those emission standards, we must scrub more of

Left: An employee turns a valve on a wellhead at The Geysers and sends the steam speeding through insulated pipes to a generating plant. (Courtesy of the Union Oil Company of California)

Below: An employee checks on the insulated pipeline that carries natural steam from the wells to the generating units. Note the archway of pipes in the rear center. This is an expansion loop. It allows for pipe contraction when the power plant is shut down, and for pipe expansion when steam rushes through the power plant at start-up. (Courtesy of the Union Oil Company of California)

that dissolved H$_2$S out of the water before it goes to the cooling towers. It's a very complicated process, and we're working on still other means to clean it up."

By 1980, the Known Geothermal Resource Area (KGRA) at The Geysers was producing 908 megawatts of electricity from the steam roaring out of more than 200 wells at its fifteen individual power plants. And, to quote Miller again, "Although most of The Geysers' power does not serve San Francisco directly, we are, nevertheless, supplying 90 percent of that city's electrical needs. And we believe that the total capacity of the fields may be as much as 2,000 megawatts, which could, roughly, serve an American city of two million inhabitants.

"Still, although The Geysers' steam wells are long lasting and may continue to produce for thirty years or more, each new plant calls for the drilling of two new wells every three years, in order to make up for those that lose pressure during plant operation."

Asked how the cost of geothermal electricity compared with the cost of electricity produced from other energy sources, Miller had this to say: "Geothermal energy is the least costly new generating source in the Pacific Gas and Electric system. However, the price of steam (the price paid to the suppliers) has been rising steadily because it's linked to the rising cost of oil and other energy sources."

Despite the rising cost of steam, a Pacific Gas and Electric press release noted some good news in May of 1980, claiming that "each 100,000 kilowatts [100 megawatts] of geothermal capacity replaces the need to burn about a million barrels of oil a year in conventional oil-fired electric generating plants."

A glance at the following figures, taken from a study by the Pacific Gas and Electric Company, will show how low the total production costs run in their geothermal facilities compared with those in their nuclear, coal, and oil generating facilities. These figures (which include the cost of drilling, construction, and installation, as the case may be, up to the point of electrical generation) prevailed in 1977 and are expressed in terms of KWH, meaning kilowatt hour—1 KWH would illuminate ten 100-watt light bulbs for one hour.

• For geothermal dry steam, the cost was 1.7 cents per KWH when the plant was operating at 80 percent of capacity.

• For nuclear, the cost was 2.3 cents per KWH when the plant was operating at 70 percent of capacity.

• For fossil fuel coal, the cost was 2.9 cents per KWH when the plant was operating at 70 percent of capacity.

• For fossil fuel oil, the cost was 4.5 cents per KWH when the plant was operating at 80 percent of capacity.

And note that no plant operates year round at 100 percent capacity because it must occasionally shut down for general maintenance or repair or cut back because of reduced demand.

Across the Pacific, right on top of the Ring of Fire, lies the Japanese Archipelago with 20,000 bubbling hot springs and countless fumaroles puffing dry steam. Some of these fumaroles have been puffing enough dry steam to tempt the research and development groups to look twice, reach for their calculators, send for the geologists, make surveys, and go into the business of drilling for steam and producing electricity.

As is true in many parts of the world, it's not too hard to find fumaroles, but drilling into them and finding abundant high-temperature steam is something else again. Nevertheless, geothermal electricity from dry steam is now being generated at Matsukawa. This is only a small power station, but the results are encouraging. And the opinion of the man in the street, hurt by the high cost of imported coal and oil (since Japan has so little of her own) is predictable. "Let's not stop with Matsukawa. Let's have a real go at our geothermal resources. We've got plenty of them."

By the 1980s the Electric Power Development Company was indeed having a go at it. A five-year plan had been put into operation, and highest priority was going to the construction of a large power plant at one of the seven geothermal sites known to have high-temperature steam. Still, although public opinion is all for the development of geothermal electricity, that same public drags its feet. The Japanese do not want any industrial tampering with their hot-spring spas or their national parks, which contain most of the country's geothermal resources.

Although the development of electricity from dry steam was a twentieth-century phenomenon, the known dry steam reservoirs are few in number. Engineers, accordingly, have been turning their attention to geothermal hot-water reservoirs, which are remarkably abundant. So the future may well hold a pleasant surprise, because the development of low-cost electricity from the earth's natural hot water may just turn out to be a twenty-first-century phenomenon.

4

In Search of Hot Water

To generate electricity from geothermal hot water, engineers need two things: appropriate technology and abundant high-temperature water. The technology is available. It's efficient and durable and already producing low-cost electricity in New Zealand, Mexico, and Indonesia. Abundant high-temperature geothermal fluids also are available because this planet has a vast hot-water circulation system underground. According to geologists, however, only under certain conditions of depth, temperature, and chemistry does it pay to drill into this system.

Immediately, three questions come to mind: Where does this water come from? What makes it hot? And how permanent is the source?

Well, the water that comes from the rain and snow seeps into the ground. Obviously, it cannot continue to soak down indefinitely. Somewhere it will reach an impermeable (impenetrable) rock layer. There it will turn aside and spread along the lines of least resistance until it comes to fractures or fissures in the rocks. Down these cracks

for Generating Electricity

it will flow, down to the next porous layer—to rocks that have tiny pores or spaces in them that fill with water. Such a layer permits water to flow through it and is called an *aquifer*.

Water is constantly moving underground through various aquifers. These may be shallow, or they may be thousands of feet deep and several miles wide. If the aquifer is deep enough, it may rest on the impermeable rock layer that is in contact with the earth's heat source—the superhot magma. Such an aquifer will, understandably, be very hot, and water traveling through it will soak up heat as if it were moving through a hot sponge. So a kind of geological sandwich occurs underground wherever there is an aquifer:

- There's an upper rock layer that is impervious.
- There's a middle rock layer that is porous.
- There's a lower rock layer that is impermeable and hot.

How then does the earth's interior heat get carried to the surface? The heat moves from the superhot magma up through the impermeable rock layers into the aquifers and heats the water there. Only when the heated water encounters some fractures leading upward can it rise and break through to the surface as hot water or steam.

Hot water always rises because it expands, becomes less dense and more buoyant. Consequently, it is replaced by the denser cold water that's seeping into the aquifer and circulating through it.

The diagram below shows how the earth's interior heat gets carried to the surface—from its source through the impermeable rock, into the water in the porous or permeable rock in the reservoir, and

Cross-section diagram of a steam reservoir. The heat is transferred from the magma to the water in the aquifer. From the aquifer, the heated water rises to the surface through cracks in the rocks. It emerges as hot springs or as fumaroles emitting steam. At The Geysers, the steam to drive turbogenerators is obtained by drilling down to the reservoir. (Pacific Gas and Electric Company)

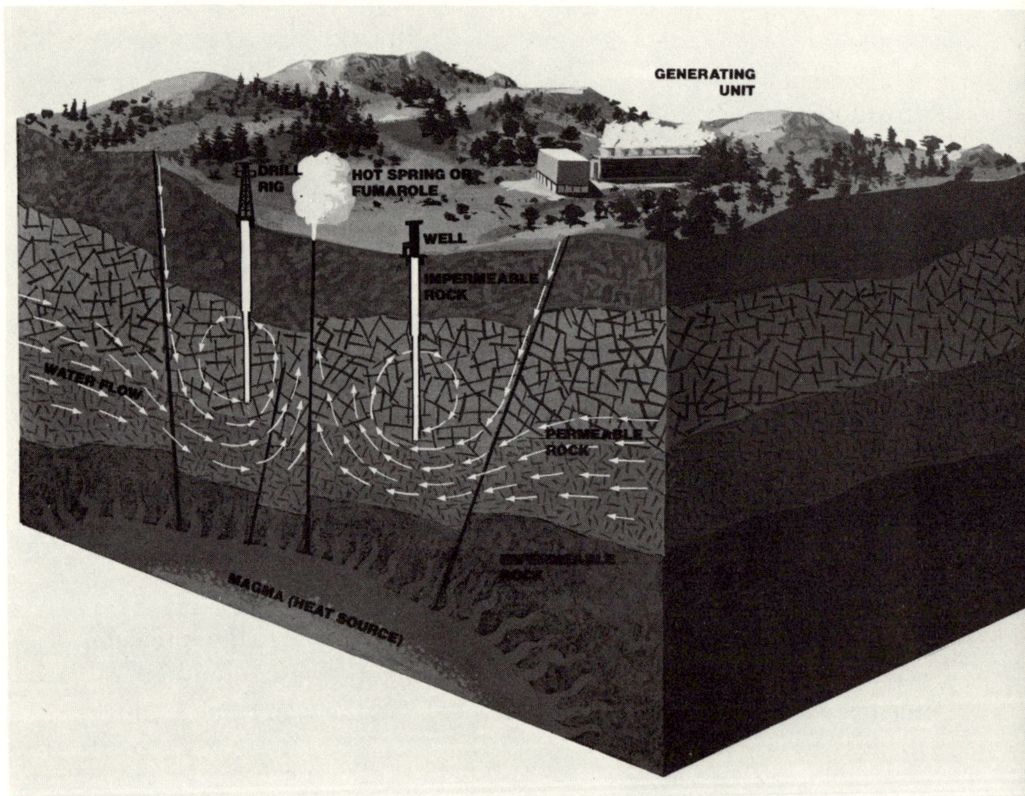

then up to the generating unit, which in this instance is the power plant at The Geysers in California. At this plant, steam is obtained by drilling 1½ to 2 miles (2 to 3 kilometers) through solid impermeable rock, down to the steam reservoir.

Geothermal reservoirs actually vary in a number of ways. Some, such as those beneath The Geysers and Larderello, contain mostly dry steam. Some contain very hot water under very great pressure. Some contain hot water at lower temperatures and lower pressures. In this chapter we will consider only reservoirs that are filled with high-temperature, high-pressure water that flashes—turns into steam—and spins a turbogenerator and produces electricity.

High-temperature hot-water fields are about twenty times as numerous as dry-steam fields, and under certain conditions such fields can be developed economically. Hot-water deposits, however, don't always announce their location by leaking through the ground in the form of hot springs or geysers. Often they are well hidden, trapped in volcanic and earthquake regions, as well as in some sedimentary areas. Prospectors know this.

In the past, a backpacking prospector would often try *dowsing* in order to locate hot-water reservoirs. He would cut a slender forked branch from a nearby tree, and holding one end of the fork in his right hand and the other in his left hand, he would slowly crisscross the field. When the free end of the branch jerked downward, that was supposed to indicate the presence of water. How hot that water might be, how far below, or how abundant—these facts the forked branch could not tell him. Now, as the twentieth century draws to a close, it is well known that if you drill deep enough, you will most probably find water almost everywhere, even beneath some deserts. That's why geologists almost explode when you mention dowsing to them. "Dowsing," they growl, "witchcraft, hokum!"

Today's prospectors operate differently. Although a few still roam the hills on their own, most of them now scout around for the government or a large company. When they find an area that looks promising, they report to their employers, who call on exploratory teams composed of geologists, geochemists, and geophysicists.

A dowser, holding a forked stick,
searches for an underground
source of water.

Although each of these groups uses different methods and techniques, they all explore the field for the same purpose· to locate hidden geothermal reservoirs and to find out about the temperature and composition of the geothermal fluids at various depths.

First, the geologists map the springs in the area (the cold as well as the hot). They record the surface temperature of the water and measure the heat flow in those springs. They also look for fractures leading down through the rocks. These too they map, as far as they can.

Then the geochemists move in to analyze the chemical composition of the spring water because this analysis will help them determine the temperature of the reservoir from which this water came.

Above: A scientist at Wairakei, New Zealand, measuring the heat flow in a hot pool. (DSIR)   *Below*: A scientist at Wairakei taking a water sample from a hot spring for chemical analysis. (DSIR)

They can do this because the chemicals that are now present in the spring were originally part of the underground rocks in the aquifer. In time, those chemicals were leached out of the rocks by the geothermal fluids down there and carried off in solution—to the underground reservoir, to and up the fissures leading to the surface. And there they are now, dissolved in the spring water. So the geochemists list the chemicals in the water; they consult tables that show the temperatures at which different chemicals dissolve out of the rocks; and they estimate the temperature in the reservoir.

Invariably, however, there's considerable difference between the estimated deep-down temperature and the surface temperature of the geothermal fluids. There are two reasons for this difference. One is that the deep, hot fluids, rising to the surface, are diluted by the cold water that is circulating through the upper rock layers. The other reason is that the upper rock layers themselves are cold, so they rob the geothermal fluids of some of their heat.

Geophysicists are the depth experts and use highly sophisticated equipment, which includes electronic and accoustical instruments.

Their most important exploratory technique, one that involves resistance to an electric current, is called *electrical resistivity*. It works on the principle that dry, hard rocks offer great resistance to electric currents, while hot water conducts electric currents with little resistance. Geophysicists, accordingly, drive electric currents into the ground through short metal stakes and measure the resistance offered to them at various depths. When a current flashes through porous rocks filled with hot water, low resistivity readings show up on the meter. When that happens, the scientists feel confident in saying that at this particular depth you may expect to find a hot-water reservoir.

Some of the other techniques used by geophysicists are borrowed from seismic (earthquake) studies. Sometimes they use a method called *seismic reflection,* in which they bounce sound waves off deeply layered rocks. Sometimes they conduct gravity studies and measure tiny changes in the gravitational pull of the earth. At other times they work with magnetics—with changes in the magnetic

Engineers measuring the electrical resistivity around a drilled steam well in El Salvador. (United Nations)

force as measured from place to place. Still another technique uses airborne scanning surveys that photograph hidden, and generally unsuspected, heat deposits. The methods used vary with conditions at the site.

Only after pre-drilling exploratory investigations are completed and mountains of favorable on-site data have been gathered will the project get a go-ahead. Then the surveyors will set up their transits and begin mapping the area. They do this right off—not only for geological purposes but also for legal reasons, because who in his right mind would want to go ahead with a geothermal project without being sure of the legal title to the land (unless it is a government project on government land). When the title is clear, contracts will be signed and big trucks will roll in carrying crew, equipment, and a diesel-powered generator to provide "juice" for the project.

Still, no matter how careful the investigators might be, the teams can never be sure of striking abundant hot water because the interpretation of the pre-drilling data might not be 100 percent accurate. Geoscientists never know for sure that they can bring in a geothermal well on a given field until after they have drilled to a reasonable depth (reasonable meaning economically feasible). As someone once said, "The drilling tool is the final arbitrator for the evaluation of underground resources."

The busy, hectic field work continues until the day when the exploratory geologists pinpoint a drilling site. Immediately, the drilling engineers in their field boots, hard hats, and protective gear, with maps and logbook in hand, call in the drilling crew. They erect a portable derrick of steel girders. They assemble the drilling machinery and tow the generator nearer to the site. They hook up the drill bit and set it into the ground. The well is now "spudded in," and the drill begins to turn, chewing its way through the cap rock—the impervious upper layer—down to the porous, water-bearing aquifer. The search is on for earth heat and a *conductor* of that heat in the form of water or steam—which will bring the heat energy up to the surface.

Sometimes the drilling ends up with a dry, nonproductive hole that produces neither hot water nor steam. But even a dry hole is not a total loss because it can add to the picture of what's underground. Sometimes, engineers have to drill a dozen holes before a "discovery well" comes on stream. Or they may drill and drill and end up with only dry holes because, on that field, the water may be too far below the surface or not hot enough or not abundant enough. To geoscientists, this is no reason for despair. They know that, in general, there's a tremendous amount of heat and hot water stored within a few miles below your feet, whether your feet are standing in your garden, on a sidewalk in New York, San Francisco, or Tokyo, on a snowfield at the North Pole, on a marsh in the Brazilian jungle, or on the sands of the Sahara. So if they drill in an unsuccessful field, they use that information to determine where more favorable sites might be located. And by a more favorable site, they mean one that will bring in a high-temperature hot-water well (350 to 690 degrees F or 177 to 366 degrees C) that will roar into existence with billowing clouds of steam.

For a high-temperature hot-water well to produce roaring, billowing clouds of steam may sound like magic, but it isn't magic at all. The tremendous pressure deep in the earth keeps the high-temperature water there in liquid form. It's only when the drill bit penetrates the reservoir that the pressure is released, and when that happens, the hot water flashes—turns to a mixture of water and steam (called *wet steam*), which wooshes up the borehole like an explosion.

Have you ever seen a cork pulled out of a champagne bottle? Immediately the bubbling liquid that had been under pressure streaks toward the ceiling! Or have you ever been in a kitchen when a pressure cooker hissing away on the stove built up too much pressure? The gauge hits the ceiling as the wet steam rushes, whistles, blasts, and splutters out of the vent. The same happens with the wet steam that explodes out of a high-temperature hot-water well that is under high pressure.

Wet steam for the commercial production of electricity was first harnessed on New Zealand's North Island. It was in the 1950s that the citizens "down under" seriously announced that if the Italians could do it at Larderello, they could do it at Wairakei.

The New Zealanders' confidence came naturally because volcanoes, hot springs, and fumaroles are almost a way of life there. That land has always been wreathed in steam clouds and mystery. One Maori legend, for example, tells how a *taniwha* (a monster that lives in deep water) was chased from the center of North Island to the Bay of Plenty, how it had to travel underground, how it had to come up here and there to reconnoiter, and how each of these places is now marked by a steaming hot-water pool. The early developers of geothermal energy in the 1950s were not much more knowledgeable about the source of those hot-water pools than the ancients had been. They were, however, aware that no two geothermal wells are exactly alike and that wet steam is different from the dry kind found at Larderello. And they soon found that handling wet steam is a complicated matter. That's because in a given quantity of wet steam, the steam may constitute as much as 90 percent of the *volume* and water only 10 percent, whereas the water may constitute approximately 90 percent of the *weight* and steam only 10 percent.

Before the New Zealand Ministry of Works and Development and Ministry of Energy would agree to spend any of the taxpayers' money on the new power technology, they wanted answers to three hard questions:

- How much steam can be gotten from the fields?
- How long will the steam last?
- Where should the drilling take place?

Naturally, the scientists hastened to search for the answers, but in the meantime the work crews had to get along on information that was limited and technology that was meager. Soon, however, by using light rigs to drill down 750 to 1,000 feet (229 to 305 meters), they found that the temperature increased with the depth. Further drilling to depths of 1,500 to 2,000 feet (457 to 610 meters) showed similar results. And still greater temperatures were found at 2,500 feet (762 meters).

Based on these and other findings, the scientists designed working models that demonstrated the details of the subterranean hot-water circulation system. These models showed that vast hot-water reservoirs were within easy reach of the newer, heavier power drills. But the geologists pointed out that most of that hot water would never come to the surface unless it was drilled for and presented with escape routes. In time, drilling crews sank a number of 2,000-foot (610-meter) wells at strategic sites. As soon, however, as the drills penetrated those 500-degree F (260-degree C) reservoirs and the pressure was released, the hot water flashed, and steam, water, rocks, sand, and pebbles blew out—exploded upward with such force that the air thundered and roared. Even though the men clutched their hard hats and raced for cover, more than one received third-degree burns from the tempest of hot water that soaked through their moisture-proof gear. Occasionally, too, a labor-

Opening a new bore by vertical discharge to clear away debris paves the way for temperature measurements and for fitting the wellhead equipment. (MWD)

er would find that his high, heavy-duty boots could not keep him from slipping in the steaming, scalding mud that sometimes bubbled around the borehole. Today such blowouts are rare because the wells are now drilled under control and opened only when the operation is considered safe.

The New Zealanders soon realized that they had an abundance of available wet steam at the Wairakei field but had no precedent to follow. Whereas the Italians had simply captured their relatively clean dry steam at the wellhead and used it to generate electricity, New Zealand's engineers found a number of problems that had to be solved before they could even begin to think about generating electricity, such as:

• The wet steam wouldn't turn the turbines smoothly because

Drilling a new steam well alongside a producing bore.
(New Zealand Consulate General, N.Y.)

    Geothermal Energy

conventional turbines are made to rotate by the pressure of dry steam only.

• The water in the wet steam was so heavily chemicalized with dissolved silica, sodium chloride, and calcium carbonate that the metal components of the installation were quickly attacked by scaling and corrosion.

What to do? It was agreed they would have to devise their own technology to separate the water from the steam. If they could do that, they would be left with relatively clean dry steam with which to run the turbines. And, with the new stainless steel alloys that the developers at The Geysers were using, the New Zealanders expected that their installations would also be relatively safe from chemical attack.

Generally speaking, however, problems are rarely solved successfully without reference to the human factor. In this instance, while the new separation technology was urgently needed so that the wet steam could be processed, the matter of noise had to be corrected first so that the men could do their work. When the steam rushed from the wells, the noise was so deafening that the workers could hardly endure it, although every one of them wore snug-fitting earplugs and earmuffs.

In short order, ingenious engineers solved both problems. They built great structures near the wells: *silencers* to reduce the noise and *cyclone separators* to remove the hot water from the steam.

The silencers are twin towers that receive the wet steam directly from the wells when it isn't required to turn the turbines. In the silencers, this wet steam (which the New Zealanders refer to as the *discharge*) is directed to pass by a steel wedge at the base of the towers. As it hits the wedge, it flows off in two separate jets. The steam swirls upward, expands rapidly, loses speed, and billows out of the top with only a dull roar. At the same time, the hot water gets thrown to one side, falls to the base of the towers, and runs off in disposal pipes.

But when the turbines call for steam because electricity is to be produced, the rushing discharge from the wells is sent directly to

Twin silencers and the cyclone
separator (*center*) in operation
at Wairakei. The processed
steam, now quite dry, speeds
from the separator, through
the pipeline, to the power plant
at the Waitako River. (NPS)

the cyclone separator. Here it enters at high speed, goes into a spin, and whirls around like a cyclone. Here, also, two things happen at the same time:

    • The water in that wet steam gets thrown to one side and down to the base. Then it is led off to the silencers (where, because the

pressure is lower, some of it boils at a lower temperature and escapes in clouds of low-pressure steam that hang over the field).

• And the high-pressure steam (now that it is relieved of its water) gets quite dry. This dry steam is collected in insulated pipes and led on a two-mile journey to the power plant.

Steam mains, two miles long, carry the steam to the power plant at the
Waitako River. (NPS)

At Wairakei, the pipelines are that long because it was considered
more economical to build the power plant at the Waitako River,
where so much cooling water would be available to condense the
steam after it is exhausted from the turbines that no cooling towers
would be needed.

Actually, the same steam is used three times by turbines of decreasing power. First, the steam goes to the high-pressure turbines
and produces electricity. Then the used steam, now much reduced
in pressure, is sent on to the intermediate-pressure turbines and
produces more electricity. After that, the twice-used steam, now
quite low in pressure, is sent on to the low-pressure turbines to produce still more electricity.

By this time, that steam, which had been traveling for two miles from the noisy, steam-shrouded field at Wairakei, has completed its work in the quiet, efficient power plant on the Waitako River. After turning the turbines, that spent steam is scrubbed by vacuum pumps for gas removal. Then it is sent to the condensers, where it turns to water, ready to be discarded. Pipes catch the discard water, and carry it to the river. All that remains of the steam that roared out of the Wairakei wells is a faint odor of a mixture of gases.

New Zealand's geothermal story is only about thirty years old. It began at Wairakei with an installed capacity of 192 megawatts, but by present estimates this is only a small fraction of the country's capacity to generate geothermal electricity.

Interior of the power plant at the Waitako River, where electricity is generated. (NPS)

Many experts consider the geothermal sources quite reliable. As A. W. Reed writes in his excellent little publication, *Power from the Earth at Wairakei,* "There are no nine-to-five working hours with geothermal steam and no seasons when the output rises and falls. The steam bores roar unceasingly, night and day, year after year."

Many optimistic New Zealanders have been known to argue that the size of the resource and the resulting impracticality of exhausting it are reason enough to consider geothermal energy a renewable resource. Nevertheless, though the steam goes on roaring year after year, the output in any given field cannot be guaranteed. Neither can any field be guaranteed to produce indefinitely. No one knows the permissible rate at which the geothermal fluids may be removed from the wells. Extract too much and you may reduce the pressure of the steam, and if this happens, there's always the danger of cold water from the surrounding fields entering the reservoir and reducing the temperature. Even if the temperature is not drastically reduced, the lowered pressure would permit some of the dissolved chemicals to come out of solution. These chemicals could, in time, plug the skin, seal off the heat supply, clog up the well, and shut it down.

It's easy to see that the work of the engineers and geoscientists is never done. They must be on constant alert for subtle happenings— for clues to possible changes in the wells, around the wells, and in the technology.

Because the New Zealanders proved so successfully that energy from hot-water wells could be converted to low-cost electricity, interest in hot-water fields around the world began to grow.

In the 1960s, Mexican engineers spudded in a number of hot-water wells at Cerro Prieto in Baja California and struck water hot enough to flash into steam. After ten years of study and development, the Cerro Prieto plant began producing 75 megawatts of electricity. Some time later, the plant increased its capacity to 150 megawatts, and plans for 1980 call for another increase that will boost generating capacity to 400 megawatts.

Field workers opening the discharge valve of a geothermal well at Ahuachapán, El Salvador. (United Nations/Yutaka Nagata)

Recently, too, other countries have been encouraged to locate, explore, and develop their fields with the technical assistance of the United Nations. Surveyors and drilling crews have, accordingly, moved into Kenya, Indonesia, Taiwan, El Salvador, Chile, the Philippines, and Turkey. In varying but still limited degrees, some of these countries are beginning to build small power stations and produce low-cost electricity from their high-temperature hot-water wells.

When the hot-water wells are of low temperature, developers know that they won't be getting wet steam roaring out of the bores. In such instances they like to think about installing either a flash system or a binary system. These systems, however, can be used only where the geothermal fluids are just barely mineralized. If the flu-

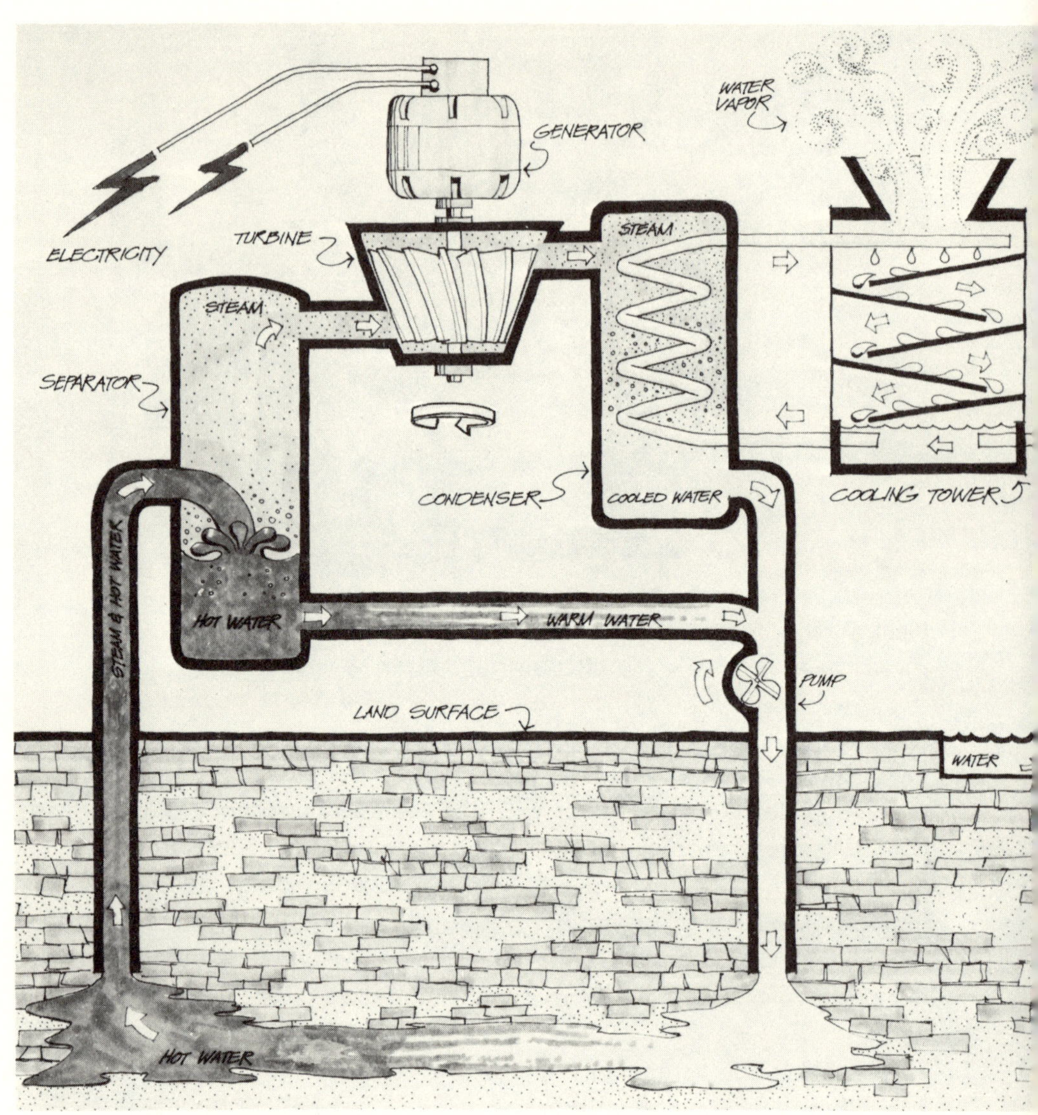

A simplified diagram of the flash system for the production of electricity. (Chevron Resources Company)

ids are too briny, the pipes and the separators will become scaled, corroded, and fouled up with deposits in no time flat.

Additionally, these systems are expensive, and since the bottom line in business invariably is "cost," few such projects have been considered seriously.

Now the dollar sign may come first with developers and investors, but research comes first with scientists. Accordingly, teams of scientists and engineers have been working on flash and binary systems for years. Always their aim has been to make these systems more technologically efficient and economically feasible, and they have made considerable progress.

To produce electricity with a flash system (see page 72), a trained crew must pipe the hot geothermal fluid up from the well to a separator. There, because the pressure is now naturally released, some of this fluid flashes into steam that rushes off to turn a turbine that spins a generator and produces electricity. The spent steam is then chilled in a condenser, whereupon it turns to water, which is pumped back into the ground. And the part of the geothermal fluid that had not flashed into steam in the separator is also pumped back into the ground. This process continues as long as the pipes keep bringing the hot geothermal fluid to the separator.

To produce electricity with a binary system (see page 74), a specialized crew must use the heat exchange method, which is altogether different in principle. Here, heat from the geothermal fluid is transferred to another liquid—a refrigerant—that vaporizes and turns the turbine. This is how it works:

• The hot geothermal fluid is continuously pumped through a pipe that runs from the well up to and through the heat exchanger, and is then pumped back into the ground.

• The heat-exchanger vessel contains a turbine fluid, a liquid refrigerant such as Freon or isobutane.

• The hot-water pipe, passing through the heat exchanger, makes the liquid refrigerant there boil and vaporize. As it vaporizes, it turns into a highly pressurized gas that flows up a pipe leading to the turbine. It spins the turbine, which turns the generator, and presto!—electricity is generated.

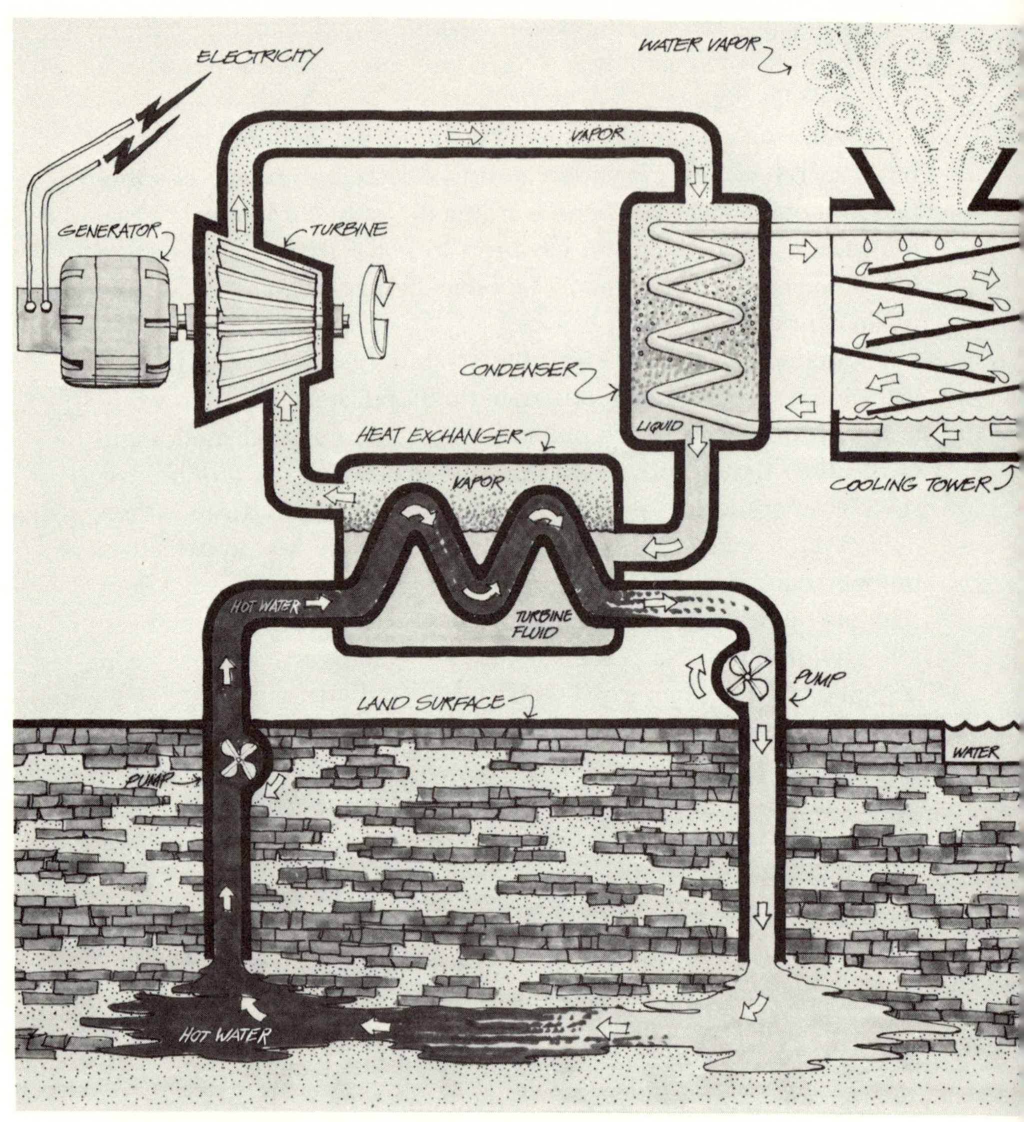

A simplified diagram of the binary system for the production of electricity. (Chevron Resources Company)

So far so good, but now the question arises: "What about the vaporized refrigerant? What happens to that?" Well, after it spins the turbine, it loses pressure and flows on to the condenser. In the condenser, it is chilled by cold water from a local source. The chilled refrigerant begins to liquify, and as a liquid, it is recycled—pumped back to the heat exchanger—and the process is repeated again and again *ad infinitum.*

In 1970, Russian engineers made history with a binary pilot project in Paratunka, Siberia. They used below-boiling geothermal fluid (176 degrees F or 80 degrees C) and produced 700 kilowatts of electricity. This was a milestone because it proved—three years prior to the fossil-fuel crisis—that the binary system works. But don't ask about the cost. It was formidable.

In 1980, a commercially sized power plant using both binary and flash systems was contracted for at Heber, California. There, the Southern California Edison Company, using geothermal fluid provided by the Chevron Resources Company, will produce 50 megawatts of power—enough to supply 50,000 people in the Imperial Valley.

By the 1980s, the United States, which is richly endowed with geothermal resources, had become the undisputed leader in research and development.

In Idaho, a small prototype binary power plant was operating on the Raft River.

In El Centro, California, engineers were experimenting with a *downhole pump* that would force low-temperature water in the 176-to-300 degree F (80-to-149 degree C) range to flash. They lowered a downhole pump into a low-temperature well. The pump was switched on. It reduced the pressure there to a point below atmospheric pressure, and the water flashed, turned into low-pressure steam, and roared out of the well. With low-pressure steam and with low-pressure turbines they were soon generating electricity!

In East Mesa, California, the Magma Power Company was building the first power plant to utilize these three technologies: binary

The Nero Geothermal Well in Hawaii is the hottest in the United States (675 degrees F or 357 degrees C). (U.S. Department of Energy, photo by John Shupe)

systems, downhole pumps, and downhole heat exchangers. This is the Magmamax, of patented design, which is expected to produce 11.2 megawatts of electricity.

Additionally, some fifty exploratory wells were being drilled annually, mainly in the Imperial Valley, East Mesa, and Salton Sea areas of California; in the Sulphurdale belt in Utah, and in Valles Caldera, New Mexico. Nor should we forget to mention the hot-water deposit at Puna, Hawaii. According to the Hawaii Institute of Geophysics, this reservoir with an estimated potential of several hundred megawatts has the hottest geothermal water in the United States and is virtually pure enough to drink.

Who would have believed a hundred years ago that electricity would be generated from hot water? And now we have all these exciting power plants and experimental projects roaring around the world and illuminating it. Of course, developers prefer the high-temperature water, but we have to remember that for every geothermal field that is very hot, there are two low-temperature fields where the water doesn't contain enough energy to flash.

In the past, when developers drilled into low-temperature fields, there was disappointment, consternation, and even bankruptcy. Now, many are discovering that low-temperature water can be used very economically for *nonelectrical* purposes—as sources for direct on-the-spot energy—as you will see in the next chapter.

5

Geothermal Energy fo

In 1974, Dr. I. M. Dvorov, high-ranking member of the U.S.S.R. Academy of Sciences, Moscow, was scheduled to speak at Klamath Falls, Oregon. At the last moment Dvorov, finding himself unable to attend, sent along a paper for inclusion in the proceedings. This paper focused on the multipurpose uses of the kind of hot water that underlies much of Russia—low-temperature water 104 to 392 degrees F (40 to 200 degrees C). His introduction, which follows, describes some of these uses so convincingly that we're right there with him in that gem of a town, "Teplogorsk." He wrote:

"The northern town of Teplogorsk situated not far from the polar circle is considered the cleanest town in the world. Despite rigorous climate, the town does without coal, oil, gas, peat and wood. Perhaps you think that the town of Teplogorsk is heated by electric energy produced by atomic power stations. Nothing of the kind. The town is heated by deep heat of the Earth. Here great amounts of thermal water and steam-water jets outflow from the Earth.

"The streets of the town are perfectly clean, the houses are light

101 Uses

colored, and pavements and roadways are not covered by thick snow despite abundant snow-falls (up to 2–3 m) [7–10 feet] in winter, as pipelines pass under them.

"A geothermal power station has been constructed in the suburbs of the town. Its turbines are put in motion by the energy supplied from the Earth's interior. Such a station does not need the production and transportation of fuel. Neither does it need cumbersome equipment consuming the fuel, boiler shops, access roads, or storage facilities for fuel, etc. A geothermal power station provides the town with the cheapest electric energy.

"Pipelines diverge into all directions from the geothermal power station. Waste, but yet warm water, its temperature being 100–110°C [212–230 degrees F], runs through the pipelines towards the town. The water heats living houses and commercial enterprises, satisfies the requirements of the population for hot water, warms street pavements and roadways, heats green-houses and swimming pools.

"When entering the town, one can see a huge white cloud in the distance: here is the central town swimming pool in which citizens swim all the year round. Besides this swimming pool there are many others in the town.

"Aside from the central swimming pool, massive glass windowed buildings are seen. This is the Central Park of Culture and Rest with evergreen trees, bushes and flowers of all kinds collected from all parts of the Globe.

"Long and wide glass covered corridors stretch in all directions from the main valley of the park. These are green-houses where two to three harvests of cucumbers, tomatoes, onions and other vegetables are gathered every year. Bananas, grapes and other subtropical cultures grow in Teplogorsk too. So, the inhabitants are provided with fresh vegetables and fruits all the year round.

"We haven't said a word about the quality of water. As it has sulphate-hydrocarbonate sodium composition, it is successfully used by inhabitants of the town for treatment of the alimentary canal, liver and other organs, as well as disturbance of metabolism. It is astonishing, but there are no hydropathic establishments in the town. If necessary, patients take a bath of a certain temperature at their homes. As the water is slightly mineralized and alkaline, it is used at home for various domestic needs.

"A reader will certainly say that such a town does not exist at all, that it is merely an imagination of the author. Indeed, we do not have such towns in our country yet, but they will appear on the Kamchatka and Kuril Island, in the regions of permafrost on the Chukotsk, in West Siberia and many other places of the USSR.

"Such towns and settlements may [also] appear in the nearest future in Alaska and other states of North America."

The "nearest future" is now, but there are still no such towns as Dvorov envisions except for certain stellar examples in Iceland, and even there, geothermal fluids are limited to nonelectrical uses. A Teplogorsk enjoying multipurpose utilization of geothermal energy is yet to become a commonplace. Worldwide, however, the idea has taken root. As we have seen, scientists have learned to convert high-temperature fluids (dry steam as well as wet) to electricity. Now they are learning to process low- and medium-temperature waters for a hundred different uses: for heating and cooling buildings, bathing, cooking, fast freezing, drying, curing, and more.

Here's how Iceland has come closest to Dvorov's dream and in so doing become the world's geothermal showcase.

Lying athwart the Arctic Circle, this little country had energy problems because it had no coal mines and no oil wells. Even the forests that originally grew there had been chopped down ages ago for firewood. To keep themselves and their economy going, the people had to find a native source of cheap and abundant energy. And they had it in the form of earth heat. To this day, the country has more than a hundred active volcanoes, countless hot springs, and warm, fast-flowing rivers tumbling down from the highlands. Actually, in small ways, they had been helping themselves to this natural earth heat for centuries, as you can see by the following little anecdote, contributed by Edward F. Wehlage, president of the International Society of Geothermal Engineers.

It seems there was once a legendary geothermal pioneer—a lusty thirteenth-century writer, Snorri Sturluson. Now Snorri was a fellow who was partial to warm beds and hot baths, which were not easy to come by in Iceland's climate that ranges from cold to not so cold (31 to 48 degrees F or minus 1 to minus 9 degrees C). Still, there was a spring on the hill behind Snorri's house and in that spring was bubbling hot water. So Snorri, who was ingenious, practical, and handy with tools, designed a central heating system for himself. With wooden pipes and shallow trenches, he ran the hot water from the spring to his house, assured himself of free hot-water comfort, and became the envy of his shivering neighbors during the long, cold Icelandic nights.

Having hot water in abundance and using it to warm a bed or a bath is not the same as utilizing it for large-scale space heating. It is true that by the time the twentieth century rolled around the Icelanders were generating enough electricity at their hydroelectric stations to light their homes and streets, but very few had warm buildings because imported coal and oil supplies were costly. With the start of the new century, however, new thinking took over. Expanded social and educational needs called for low-cost heating so that the working day, which traditionally had been limited during the six months of winter, could be extended to eight hours all year round; and so that night schools, lectures, and social affairs could be held in the evenings. Therefore, the resourceful Icelanders, having plenty of low- and medium-temperature hot water, and plenty of low-cost hydroelectricity with which to run their pumps, turned to the geothermal space-heating business.

Development began in 1928. Drilling rigs were hauled to the outskirts of Thvottalaugar, to a hot spring that was used for washing

At the turn of the century, the people used to come to this hot spring at Thvottalaugar, Iceland, to do their laundry. (Embassy of Iceland, Washington, D.C.)

clothes. A well was spudded in, and the engineers soon found that the deeper the drill bit churned, the hotter the water became till it registered 183 degrees F (84 degrees C). At this temperature it wasn't hot enough for electrical generation, but it was more than adequate for heating purposes. So it was captured in insulated pipes and sent on a 2-mile (3-kilometer) trip to heat the town's homes (about seventy of them), and the schools, hospital, and new swimming pool. What excitement this project created! Naturally the drilling rigs were quickly hauled from one town site to another, wherever hot springs simmered in nearby pastures. Invariably, the new wells were quite shallow (about 630 feet or 192 meters), and just as invariably, the water was of good drinking quality and surprisingly abundant. In Reykjavik, for example, the development grew to such an extent that the water was pumped to surface storage tanks and then piped to designated points in the city, where it was metered for billing. Why billing when geothermal water is itself free? The answer is: to cover the cost of drilling, storage, and transport.

Presently 97 percent of the buildings in Reykjavik are billed for the low-cost geothermal water that heats them. Oil and coal are no longer imported for that purpose because there's no need. And far from being smothered under overhanging clouds of black chimney smoke, Reykjavik (which in Icelandic means Smoky City) is now the cleanest city in the world. Paint on the houses lasts longer, fire hazards are reduced, and insurance premiums are lower.

And if you think that Reykjavik's 97 percent geothermal heating record is good, look to the town of Hveragerdi because it is solely heated by geothermal hot water.

Now this low-cost hot water was too good a thing to be limited to heating homes, so the enterprising Icelanders soon expanded their use of this geothermal bonanza. Some set up facilities to dry tons of produce, seaweed, grain, and fish. Others built wool-processing shacks, where workers first washed the wool in the hot water to remove animal fat and dirt, then dried it with geothermal heat. And many householders built snug little greenhouses and grew vegeta-

Near Reykjavik, Iceland, the children use the geothermal hot-water pipe as a bicycle track. (United Nations)

Tomatoes and other vegetables (as well as flowers) thrive year round in Reykjavik greenhouses. (© National Geographic Society, photo by Emory Kristof)

bles, flowers, and fruit—even grapes and bananas—during the severest winters.

Although low-temperature reservoirs are plentiful in the earth's underground circulation system, the Icelandic way of drilling shallow wells and piping the hot water to the towns won't work everywhere. In the Alaskan interior, for example, the population is so sparse and the annual income so low that a drilled well for a lone house or two is financially out of the question. Still, the people need energy. Heating oil, which must be transported over great distances on virtually nonexistent roads, is formidably expensive. And yet, there are hot springs bubbling all over the terrain.

Within the last few years, the Alaskan government has been encouraging studies to find out how this hot-spring water might be utilized for heating purposes. Now, in a comprehensive 1976 publication, *Geothermal Energy and Wind Power*, sponsored by the

Icelandic children showing off bananas that were grown in the greenhouse. (Mats Icelandic Photo and Press Service)

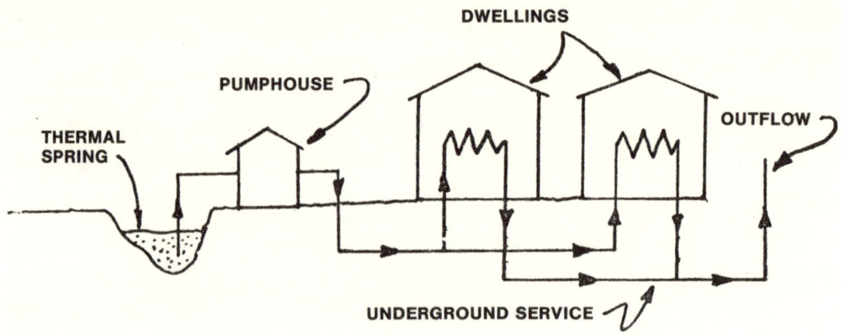

A primary space heating system in Alaska. (From *Geothermal Energy and Wind Power . . . Alternate Energy Sources for Alaska,* by permission of William McConkey)

Alaska Energy Office, the editor Robert B. Forbes, top geothermal scientist, offers two proposals. By use of a simplified design (see the diagram above), he shows a system planned to lift hot water from a spring to a pumphouse. From the pumphouse, this water is circulated through an underground pipe to the radiators in the dwellings, and then back underground to an outflow pipe.

If, however, the geothermal water should happen to contain a concentration of dissolved salts and other minerals, there are problems. As the water cools in the pipes, the salts are sure to precipitate—come out of solution—and clog and corrode the pipes. For such a contingency, Forbes offers another design, an interesting scheme that would take care not only of these problems but also of the possibility of frozen pipes. (Alaskan winter temperatures drop to minus 49 degrees F or minus 45 degrees C.) This plan uses a *surface heat exchanger* consisting of two facing coils. (See the diagram on page 89.) The hot-spring water would *not* be circulated through the dwellings. This may sound strange when we're talking about a geothermal heating system, but it is really quite logical. The spring water is pumped directly from the spring to the first coil (which is the first half of the heat exchanger) and then directly back to the spring.

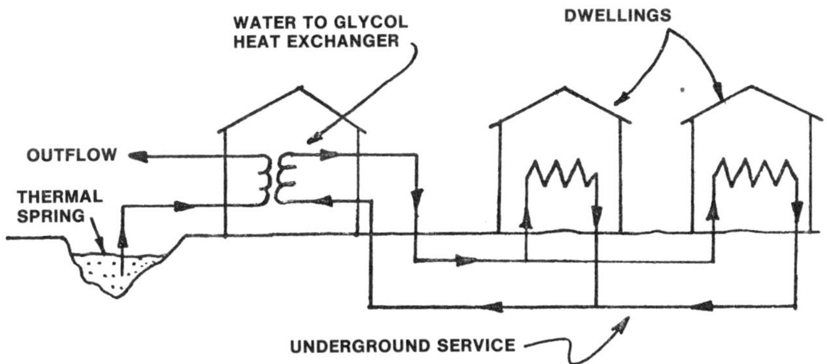

A simple binary space heating system using glycol. (From *Geothermal Energy and Wind Power . . . Alternate Energy Sources for* Alaska, by permission of William McConkey)

Now note that as this first coil is heated by the geothermal water flowing through it, the second coil, being so close by, also heats up. This second coil is filled with glycol, a form of antifreeze. The hotter the glycol gets, the faster it circulates: through the attached pipes that run through the buildings, through the radiators, and back underground to the coil. This is obviously a closed set-up, and it goes on endlessly, as long as it is kept heated by the first coil. It requires no great technology and very little maintenance because the glycol is easy on the pipes and it won't freeze.

Wherever geothermal fluids are available, they are, as often as not, heavily mineralized. Alaskan scientists are trying to meet this challenge with surface heat exchangers. Elsewhere, engineers who work with drilled wells sometimes suspend *downhole heat exchangers* right inside the well.

Klamath Falls, Oregon, is one of the places where downhole heat exchangers are commonly used. Here some 400 wells, with temperatures close to boiling, serve a number of public buildings, hospitals, schools, churches, homes, apartment complexes, and business establishments.

There are many kinds of downhole heat exchangers. Here's how one of them functions. (See the diagram below.)

This heat exchanger is an arrangement of pipes lowered into the hot geothermal fluid in the well. The pipes are filled with fresh cold city water. A small electric pump is turned on to start the water flowing through the pipes. In a short time, as the water heats up and expands, natural circulation-flow takes over. Then the pump is no longer needed, so it is switched off.

In this installation, the circulation is clockwise. The fresh water moves continuously, down one leg of the system, up through the second, through the radiators in the building, and back to the first leg in the well. Occasionally, the pump may be needed briefly to help the heated water flow more vigorously.

After looking at the diagram, you may wonder about the paraffin seal sitting on top of the water in the well. "Such a seal," say experienced geothermal engineers, "keeps the air out of the water and reduces corrosion." They also point out that downhole exchangers, which are filled with fresh water, can be used in the briniest wells

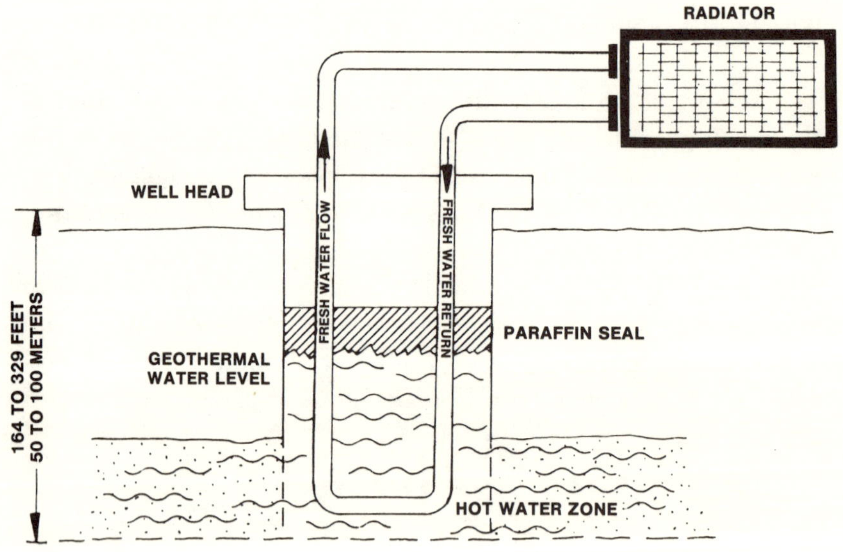

A downhole heat exchanger used in direct heat applications. (From "Technology of Utilizing Geothermal Energy" by Stanley H. Ward)

without damaging the heating system. That's because the geothermal fluids never leave the well. They never get into the heat exchanger; consequently, they never get into the heating system.

How long do such devices last? In Klamath Falls, the average life of a downhole heat exchanger is fourteen years. After that it has to be pulled up and replaced.

This chapter is titled bravely, "Geothermal Energy for 101 Uses," and yet most of the discussion has been about ways of getting low-temperature water out of the ground in order to heat buildings. In temperate and frigid climates, comfort does take high priority, so that's where the emphasis generally is. However, after people learn how to heat their buildings, they turn to learning how to cool them. Then refrigeration is just a step away and opens up all kinds of opportunities that invariably boost living standards. In the remote coastal areas of Africa and Alaska, for example, where commercial fishing is the only industry, refrigerators—and freezers, for that matter—can prevent spoilage of the catch. What an opportunity that offers to impoverished people!

Today, geothermal fluids, especially those in the low-temperature range that require no million-dollar installations, really are being put to 101 uses. Many of these were enthusiastically reviewed in Montreal, Canada, by specialists who flew in from distant lands to attend a worldwide conference—the 1979 ten-day International Conference on Long-Term Energy Resources, sponsored by the United Nations Institute for Training and Research (UNITAR), and distinguished by up-to-the-minute brilliant reports and papers on some forty alternate forms of energy.

One paper, submitted by Iceland's Dr. Baldur Lindal, a foremost geoscientist, described low-temperature water as an economic treasure for fish farmers, agricultural farmers, homeowners, miners, and spa developers around the world. (These and other uses are briefly summarized in the table on page 92.)

Other geoscientists reported on specific countries where the utilization of low-temperature water had made a marked difference in the economy. Hungary was a case in point. With 2.5 million square

SOME DIRECT USES OF LOW-TEMPERATURE HOT WATER

Approximate hot-water temperature requirements for certain uses
Baldur Lindal—Virkir Consulting Group, Reykjavik, Iceland

°F	°C	
392	200	
374	190	
356	180	Evaporation of highly concentrated solutions Refrigeration by ammonia absorption Digestion in paper pulp (Kraft)
338	170	Heavy water via hydrogen sulfide process Drying of diatomacious earth
320	160	Drying of fish meal Drying of timber
302	150	Alumina via Bayer's process
284	140	Drying farm products at high rates Canning of food
266	130	Evaporation in sugar refining Extraction of salts by evaporation and crystallization Fresh water by distillation
248	120	Most multi-effect evaporation—concentration of saline solution
230	110	Drying and curing of light aggregate cement slabs
212	100	Drying of organic materials, seaweeds, grass, vegetables, etc. Washing and drying of wool
194	90	Drying of stock fish Intense de-icing operations
176	80	Space-heating (buildings and greenhouses)
158	70	Refrigeration (lower temperature limit)
140	60	Animal husbandry Greenhouses by combined space and hotbed heating
122	50	Mushroom growing Balneology (therapeutic bathing)
104	40	Soil warming
86	30	Swimming pools, biodegradation, fermentations Warm water for year-round mining in cold climates De-icing
68	20	Hatching of fish Fish and reptile farming: salmon, trout, eel, catfish, crocodiles

Saturated steam

Hot water

Conventional power production

yards (almost 2.9 million square meters) of geothermally heated greenhouses, this country was supplying the European market with tomatoes, cucumbers, paprika, roses, and carnations.

The importance of geothermally managed greenhouses was underscored by Dr. Joseph Barnea, who chaired the conference. "Here," he said, "is a partial answer to the world's food shortage, which can be minimized by greenhouse farming." And indeed, this statement bears consideration because, in the controlled environment of greenhouses (where there are no frosts, no droughts, and no pests), farmers are already harvesting as many as three crops a year. Some are even doubling their crop yields by enriching the air with carbon dioxide that's all too ready to bubble out of the geothermal fluids.

Also noted at that conference were: aquaculture developments in Japan, where carp, eels, prawns, and crocodiles are being raised commercially in geothermally heated ponds; lumber drying kilns in Taiwan; experimental ventures with low-temperature hot water in Turkey and the Himalayas, particularly in lumber kilns and greenhouses; and the multiple uses of this energy source in the United States ranging from ground warming in parking lots to gasohol production.

Mentioned, too, was the growth of hot-water facilities in Russian towns around the Black and Caspian seas, as well as on the Kamchatka Peninsula, for therapeutic bathing, for heating homes, and for growing food. Additionally, a number of uses in the far northeast were noted: the thawing of frozen fields so that farmers could start their spring planting earlier, and the de-icing of roads so that transport could continue despite falling snow. Reports also told about the utilization of hot water in the mining of placer deposits, in adding moisture to the air in mines, and in heating concentrates in ore mills. Furthermore, since much of the Russian geothermal water is of a quality good enough for drinking when cooled, the report stated that it is being bottled and sold for medicinal purposes in some twenty-one areas.

One of the most significant contributions coming out of the Montreal conference was that made by Barnea, who, commenting on the low cost of geothermal energy in comparison with conventional forms of energy, said, "Geothermal production costs could be lowered significantly if we extracted as much heat as is feasible from every ton of steam and hot water that comes out of the wells. We need to keep on extracting that heat down to the last economic calorie." (A calorie is a unit of heat that raises the temperature of 1 gram of water 1 degree C. And a gram is equal to $\frac{1}{28}$ ounce.)

Barnea—geothermal guru, economist, and Senior Fellow on UNITAR's Project for the Future—is the No. 1 promoter of the utilization of the earth's heat. He organizes geothermal conferences, conventions, and meetings on a global scale, and invariably he gives the highest priority to what he calls the *sequential use of heat* in steam and hot water.

"What I mean," he told this author, "is that we can and we should extract this heat not once but in stages, each stage requiring heat of decreasing temperature." So, if a steam well is under consideration, he would extract this heat at least three times, as is done at Wairakei (New Zealand) with high-pressure, intermediate-pressure, and low-pressure turbines. And he urges all developers to think of what's left there after the last puff of steam is liquefied by the condensers. "Boiling brine is left, and this the New Zealanders run into the Waitako River. What a wasteful way to handle valuable heat!" Elsewhere they're even more wasteful. At The Geysers, after using the steam just once for electrical generation, they inject the leftover brine into the ground. At Klamath Falls, they barely tap the heat in the wells, withdrawing only a fraction of it to warm the fresh water in the heat exchangers. In virtually every geothermal facility today, surplus heat that could be used to good advantage is either wasted or ignored.

Barnea would utilize waste heat in multipurpose geothermal plants. A cluster of high-temperature wells in a multipurpose facility could, he explains, run an agro-industrial complex. And if, in such a complex, the steam and hot water were used sequentially, the following could be provided economically: electricity, for illumination, industrial processing, and refrigeration; heat, for drying and curing; hot water, for heating; cold water, for irrigation and in some instances for drinking. Chemical fertilizer, too, could be produced and used on farms irrigated with geothermal water pumped to the fields by geothermal electricity.

What we're talking about is the efficient use of geothermal heat, and who would want to quarrel with efficiency when that would mean money in the bank for developers and consumers. So, slowly, slowly, we catch a glimmer of a breakthrough here and there.

In Mexico the managers of Cerro Prieto in Baja California have taken the first steps in sequential use by converting their power plant to a dual-purpose facility. As of 1980, they are not only generating electricity from wet steam, but are also utilizing the separated brine in the production of fertilizer. The brine, being rich in potassium chloride and other minerals, is piped to nearby evaporation ponds in the desert, after which it is processed back at the plant and then trucked to the Mexican farmfields.

At the Raft River in Idaho, geoscientists at a series of U.S. Department of Energy sponsored projects are exploring the fuller potential of sequential—or as it is sometimes called, *cascading*—heat. Starting with geothermal fluids at 275 degrees F (135 degrees C), they first generate electricity, then utilize the exhausted steam to warm a pond growing carp, catfish, and shrimp, to heat a commercial greenhouse, and to warm crop acreage to extend the growing season. After that, the water is injected into the ground at a substantially lower temperature.

In New Zealand, the engineers at Kawerau are directing the low-pressure steam that's exhausted from their 10-megawatt power plant to serve a thriving pulp and paper facility.

By the last quarter of the twentieth century, therefore, although there was considerable talk about the sequential uses of earth heat, there was still very little to show for it. One plan from the Azores however, was especially interesting. It was a multipurpose plan devised by the team of Meidav and Meidav of Berkeley, California. Dr. Tsvi Meidav is a consulting geothermal specialist—a soft-spoken giant with a slight Middle European accent, who looks, even in dress clothes, as if he were loping across a steaming hot-water field. Mae Meidav, who joined her husband in developing this plan, is a soci-

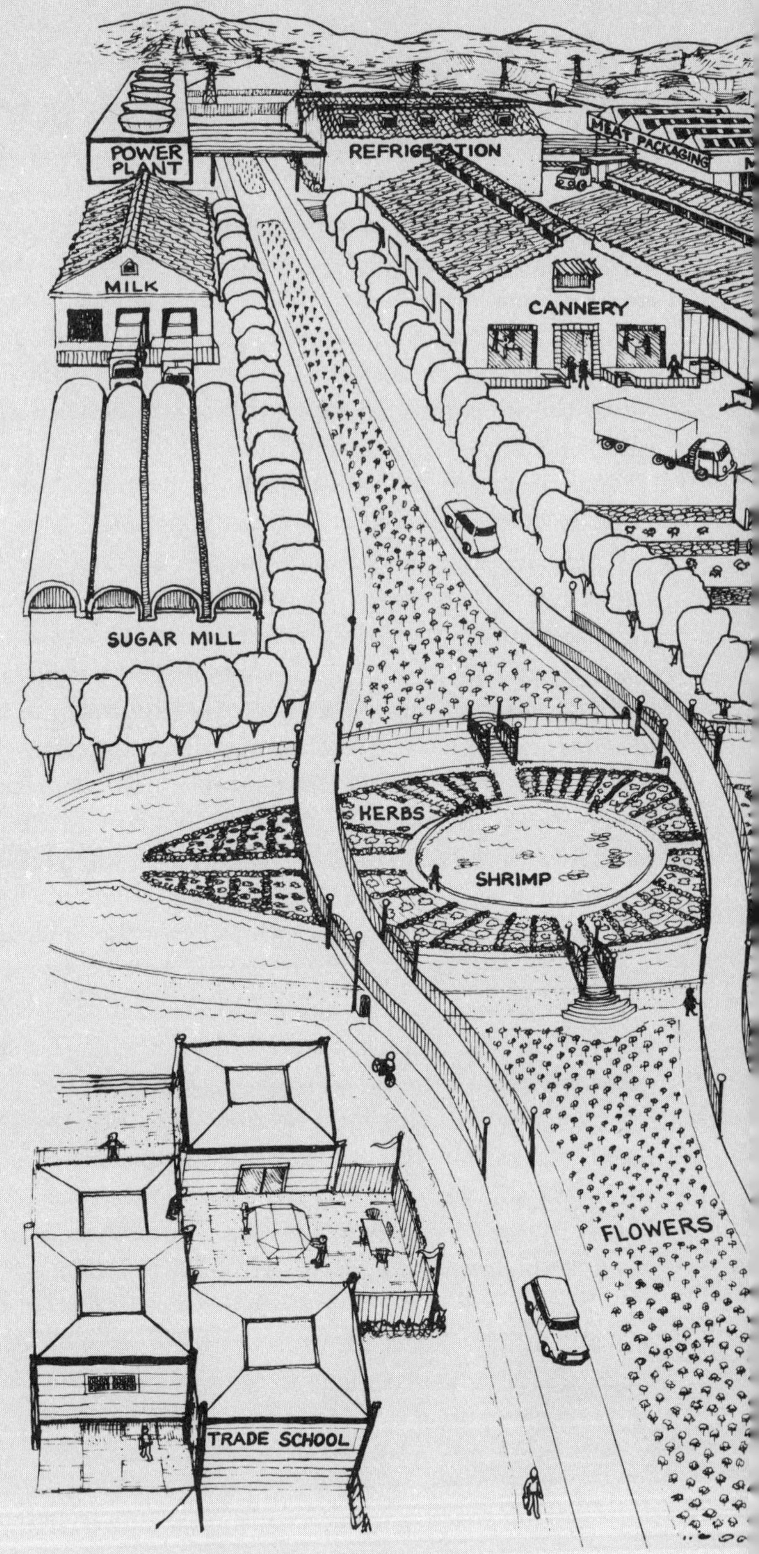

A conceptual design of the integrated geothermal energy park at Ribiera Grande on São Miguel, the Azores. (Conceived by Tsvi and Mae Meidav. Drawn by Cliff Morse)

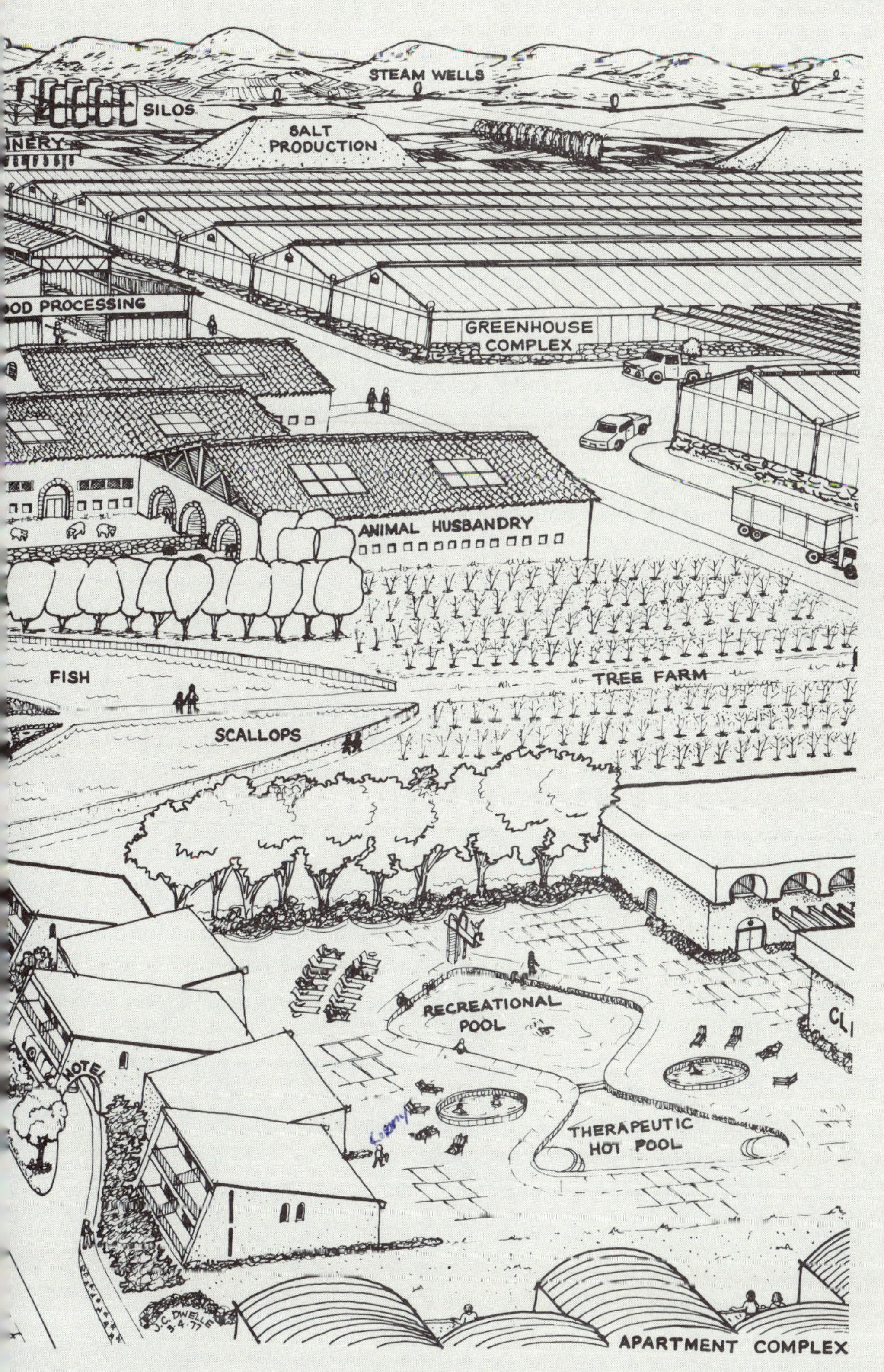

STEAM WELLS
SILOS
INERY
SALT PRODUCTION
OD PROCESSING
GREENHOUSE COMPLEX
ANIMAL HUSBANDRY
FISH
TREE FARM
SCALLOPS
RECREATIONAL POOL
CL
HOTEL
THERAPEUTIC HOT POOL
J.C. DWELLE
APARTMENT COMPLEX

ologist concerned with the human and economic aspects of geothermal systems.

Based on their 1976–1977 study of the hot-water field on the peaceful, pastoral island of San Miguel in the Azores, the Meidavs designed a *geothermal energy park* that would boost the income of the people living there.

This energy park would be located in the center of the island at Ribiera Grande. It would be an integrated complex, featuring agricultural, agro-industrial, medical, and training facilities. As Tsvi Meidav explained to a group of geothermal enthusiasts over lunch in Montreal, "Energy in this Azorean park would be utilized in such a sequential manner that decreasingly lower levels of temperature would be used in cascading fashion." He also explained that the park would be laid out to take into account the location of the field and the direction of the prevailing winds "because there's always bound to be a whiff of that pesky hydrogen sulfide, no matter how efficient the gas scrubbers might be."

There would be no lack of energy in the Meidav park, because geochemical data show the reservoir temperature to be a good 393 degrees F (201 degrees C), and Meidav's calculations indicate that the temperature, pressure, and heat flow of this resource compute to something like 200 to 400 megawatts for about thirty years. What a heartening potential this is for a little island of 200,000 inhabitants, where, presently, the total installed electric power is less than 30 megawatts.

To bring that geothermal energy to the park, drilling crews would sink boreholes, and the engineers, knowing that the water would flash into wet steam, would be prepared with separators to remove the steam from the brine. The steam would be led off to spin turbogenerators and produce electricity in a 15-megawatt power plant. The briny water, now down to about 300 degrees F (145 degrees C) would be piped from one facility to another. As long as the water registered 212 degrees F (100 degrees C) or more, it would be used for such jobs as refrigeration, sugar refining, wool processing, and the dehydration of fruits and vegetables. Low-grade heat would be piped to other facilities as the layout of this park shows. (See the illustration on pages 96–97.)

Here's how this geothermal energy on the island of San Miguel would be utilized for the benefit of the people:

At present, any milk that's surplus is converted to milk powder or yellow cheese. With low-cost electricity (or dry ice produced from the carbon dioxide in the geothermal fluids) the dairymen would be able to process a variety of milk products without fear of spoilage.

Again the catch of ocean fish is presently limited because of fear of spoilage. Here, too, refrigeration or dry ice would open the door to large-scale fishing.

Additionally, although the climate on San Miguel is quite good, it isn't good enough to permit the growing of flowers, winter vegetables, pineapples, and bananas for the European market. In geothermally heated greenhouses, however, this could be done to the tune of $8 million annually.

Furthermore, let us not forget the financial possibilities in the dehydration of pineapples, in the development of shrimp farms and sugar refineries (they raise a lot of sugar beets there), and in spas for tourists.

According to the Meidavs, a geothermally directed future looks bright for this little island.

The future—the future. Isn't it tremendous to envision integrated geothermal complexes with power plants and factories and heating centers, and mile-long greenhouses on thousand-acre plantations? And they surely will be seen before the turn of this century, because geothermal energy that is sequentially used, down to the last economic calorie, is too good an idea and too enticing a business prospect not to be explored.

6

Old Energy... New

The earth's natural heat is as old as the earth itself. Only the technology to bring it out of the ground and convert it to electrical and nonelectrical uses is new. This technology is so new it even has a vocabulary of its own—a geothermal vocabulary. If you look in old desk dictionaries, you'll not find such words as borehole, aquifer, binary cycles, energy conversion, or steam wells. Neither would you see such words as silencers, downhole pumps, downhole heat exchangers, and geothermal brines. You may find some of them now in an up-to-the-minute dictionary, but I'd be willing to bet you'd not find these newest terms: *geopressured zones* and *hot dry rocks,* which have recently been recognized as sources of ancient earth heat.

Geopressured zones are deposits of water trapped under thousands of feet of rock and clay. This water is very old, perhaps a million years or more. It is under abnormally high pressure, and is hot—with temperatures at times as high as 565 degrees F (296 degrees C). In these zones, which generally lie some 2 to 5 miles (3 to 8 kilometers) below the surface, the heat is trapped and insulated by encircling layers of sand, clay, and shale.

Technology

Before the middle of the twentieth century, few scientists had even suspected the existence of such zones. Then, oil drillers began complaining about high-pressure, nuisance deposits of hot water. "We find them," they said, "in almost all the countries where we have to drill deep for petroleum, even under the deserts." Recent exploratory reports show this to be true. In the United States alone, there are numerous geopressured zones along the shores of Texas and Louisiana and beneath their offshore waters.

In the 1970s two geopressured wells were brought on stream by the United States Department of Energy (DOE) teams. In one instance, the Chevron Oil Company was routinely drilling for gas in Louisiana when a noisy, clattering torrent of steam, hot water, and rock fragments exploded into the air with a scalding temperature of 420 degrees F (216 degrees C). The drills had struck a geopressured zone, and this immediately presented technical problems. Furthermore, Chevron was not interested in hot water. The company was in the gas production business and had a ready market for gas. It saw no profit in the exploration of a hot-water source that was ringed about with conflicting legislative acts. However, the Energy and Research Development Administration (ERDA—the predecessor of DOE) suspected that geopressured zones just might turn out to be important sources of energy. So ERDA pushed for further drilling, and a team sent the drills churning down to a depth of 19,000 feet (5,791 meters).

What happened? The hourly flow of water through an opening that was only $\frac{1}{16}$ inch ($\frac{1}{6}$ centimeter) in diameter was a surprising 100 barrels (approximately 11,617 liters), with each barrel containing an astonishing 107 cubic feet of pure methane gas in solution. And 107 cubic feet of methane, which is a form of natural gas, can heat up enough water for ten showers. Think of that. Additionally, that geopressured water was of very good quality, only one-third as salty as seawater. As one expert pointed out, "In desert countries, it could be a significant source for drinking water."

In the other instance, another ERDA team took over a small abandoned gas well, also in Louisiana. Here, the crew drilled down some 16,000 feet (4,877 meters) before striking geopressured water. In this, Delacambre Well No. 1, the water temperature was lower and the methane content was also lower, due no doubt to the fact that this well was not as deep as Chevron's and also that every well is different.

Today, informed scientists are familiar with geopressured zones and their potential, but in 1975 Professor Paul Jones, an employee of the United States Geological Survey (USGS), made history when he read his now classic paper at the ERDA-sponsored First Geopressured Geothermal Conference. In this paper, he called attention to the dual potential of geopressured zones as sources of both heat and methane. Other eminent scientists, however, see little reason for sinking huge sums of money into research and development in this energy sector.

As geologist David N. Anderson, executive director of the Geothermal Resources Council in Davis, California, told this author in a telephone interview: "This is an area that holds the least promise for us because geopressured zones are so deep, so hard to reach, and so expensive to develop. Hot dry rock research does hold a great deal of promise; however, much work must be done in this area before this source of energy will become economic."

Hot dry rocks constitute an altogether different source of geothermal energy. By definition, hot dry rocks are naturally heated *unmelted* crustal rocks. Theoretically, they can be found underfoot just about anywhere, although there is no geological evidence as yet to indicate that this resource is actually so widespread. In reality, hot dry rocks are buried deep.

A few miles down, hot dry rock temperatures hover around 350 degrees F (177 degrees C). At a depth of many miles, the heat may increase to 1,400 degrees F (760 degrees C). Within the upper six miles, according to many scientists, the hot dry rocks hold more heat than all the dry steam and hot-water sources put together.

A tremendous amount of heat is stored in those rocks, but don't expect to see hot springs bubbling up from *them* or *geysers* exploding. Those rocks may be that hot, but they are bone dry. They have to be bone dry because they are so crystalline, so impervious, and so dense that no water can circulate through them. And with no water circulating, there is no transport medium to carry the heat to the surface. That tremendous heat stays down there unless and until it is coaxed up technologically. Why don't we coax it up then? Because the cost of present-day technology (which in this area is still quite primitive) makes this form of energy too expensive as yet.

Still, the scientists aren't napping. Here's what a special team at the Los Alamos Scientific Laboratory (LASL) has been doing at Fenton Hill near Valles Caldera in New Mexico. To coax some of that heat to the surface, the men started with the idea of a closed-loop system—consisting of two boreholes interlinked at a depth of two or three miles. The scenario called for cold water to be pumped down the first borehole. There the water would be heated naturally, move on to the second borehole, and rise up it as hot water or steam. (See the diagram on page 104.)

So far so good. Engineers knew how to drill holes. But how to link two of them at that depth? How were they going to fracture the hot dry rocks down there to provide a link so that the water would

An artist's conception of a hot dry rock system whereby cold water is pumped down one borehole and rises in the second as hot water or steam that may be used to generate electricity. (From "Hot Dry Rock Geothermal Resource Ownership and the Law" by John J. McNamara and E. L. Kaufman)

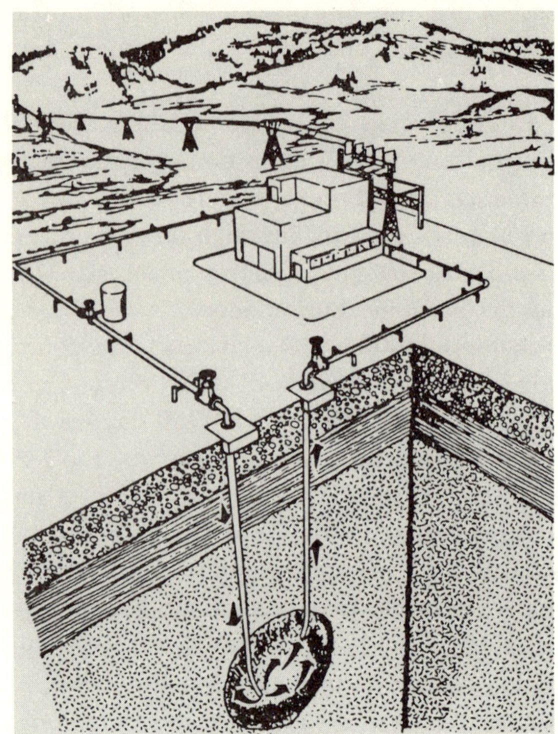

move from the first borehole to the second? Someone suggested a small nuclear device. "Drill a borehole, toss down the device, and you'll have all the fracturing you want down there." But the consensus was against this idea. The vote went to hydrofracturing techniques, the kind that oil drillers use. They drill a hole, pump down water under great pressure, and wait for the pressurized water to force the hot rocks below to crack, fracture, and break up.

That was what the LASL team did, and it worked. They got their fractured link down there, then sank a second borehole to intersect it, and they were in business.

Now, they hoped, they could begin to coax some of that heat up from the hot dry rocks. The pumps at Well No. 1 were switched on. The cold, pressurized water poured down. It was heated in the link, and nothing happened. For ten hours, nothing happened, although the pumps kept working. The crew pumped for another five hours—six, seven, eight—nothing. Another hour—thus bringing the total to nineteen hours—and still nothing. Then the clock hand moved to twenty, and everything broke loose at Well No. 2. Pebbles, steam, hot water, and rocks exploded into the air. The hidden hot dry rocks had finally, and for the first time in history, given up some of their heat—to the pumped-down water in that man-made circulatory system in which cold water goes down one well and billows up as steam from the second.

According to 1979 LASL reports, this experiment is very promising. The engineers, plumbers, and field crew have installed a system of pipes and circulating pumps, and a water-to-air heat exchanger that circulates the water under pressure. This controls flashing as well as the deposition of minerals. And the really good news is that they have managed three separate runs totaling 2,800 hours of pumping and have brought up hot water and steam. The downhole temperatures, however, were not all that startling, since the drilling had been limited to 10,000 feet (3,048 meters). In the first well, the temperature was 392 degrees F (200 degrees C). In the second, which was slightly deeper, it was 401 degrees F (205 degrees C). At the surface, the discharge registered 210 degrees F (99 degrees C), which is too low for the production of electricity unless a binary system is used, and that is still too expensive.

Although this Fenton Hill hot dry rock project is a fantastic venture in engineering science, electrical generation from this source is still far in the future. The reason is that new and perhaps totally different drilling technology that is economical has yet to be invented. In the meantime, though, no one should sneeze at a surface temperature of 210 degrees F because it could well be used for many nonelectrical purposes.

Of course more runs are being planned by the LASL teams because who could resist the spur to research, development, and invention when the United States Geological Survey estimates that in the United States alone there's enough heat in those rocks to provide the country with 1,000 times its electrical needs for centuries—if only we could learn how to reduce the cost of drilling.

So here are two new deep-drilling possibilities for the extraction of geothermal energy. The geopressured zones are already full of water that's hot and under great pressure, with an added bonus of dissolved methane gas, but the hot dry rocks promise much higher temperatures.

A still more extraordinary source of earth heat is available to us in that ultimate of heat systems, volcanoes.

Now there are two sides to everything. Until recently, a volcano was considered a source of total destruction, but the ingenious Icelanders showed the world how volcanic heat can be utilized for comfort. This they did in 1973, after a major eruption on the island of Heimaey. Almost immediately, the survivors rebuilt their little fishing town. They turned disaster into positive thinking. They looked around and saw all those smoking little mounds and banks and ridges of lava that were still too hot for the children to play on. It was hotter still some 6 feet (2 meters) below the surface, as hot as 750 to 930 degrees F (399 to 499 degrees C). Something, they thought, could be done about that. But what? After many discussions and an equal number of tries, some successful and some not, the fishermen constructed vertical shafts and positioned them in the still hot lava. The scalding volcanic gases curled up through the shafts, were caught in pipes, and led off to the town's heating system. The hot gases heated the water in the system to 176 degrees F (80 degrees C). They provided low-cost heat for homes and greenhouses. Financially, the Heimaey experiment was, and continues to be, a success. Scientifically, it is even more of a success because it marks a geothermal first in the utilization of volcanic vapors.

More recently, teams at the Sandia National Laboratories in Albuquerque, New Mexico, have been working on long-range projects for harnessing volcanic heat. These teams are part of the Magma Energy Research Project. They have plans but no deadline because extracting molten magma is a formidable affair. Imagine drilling holes into rocks that cap that hot magma, then moving into the magma chamber itself. Imagine installing heat exchangers in those holes. And imagine hooking those heat exchangers up to surface

Above: The volcano explodes behind the town of Heimaey, Iceland.
(Consulate General of Iceland, New York)   *Below:* Heimaey townsmen
brushing down their walls and shoveling away the cinders that had fallen
on their roofs and lawns. (Consulate General of Iceland, New York)

power plants. Such a scenario simply boggles the mind.

And the mind wants to know:

• What kind of metal would the drills and other technological material have to be made of to withstand the stress, the corrosion, and temperatures on the order of 1,800 degrees F (982 degrees C)?

• Would the holes drilled into the magma stay open so heat exchangers could be installed?

According to a 1979 report from the Sandia Laboratories, these two questions are being studied and the answers being tested, and the results are exciting and encouraging. The report, which was written by scientists John L. Colp and Harold M. Stoller, states very cautiously that under certain experimental laboratory conditions drills and other tools made of nickel and cobalt alloys can be expected to survive the magma environment; holes that are drilled into rocks, at near magma temperatures and pressures, show a possibility of remaining open; and thermal heat exchangers can survive while immersed in molten rock and can accomplish significant rates of heat transfer to an internal fluid.

In 1980, these findings led Dr. Colp and other team members to Hawaii to field-test some of their theories by drilling into a lava lake containing molten rock.

Their purpose? Ultimately to extract two forms of volcanic energy: heat and fuel gases such as hydrogen, carbon monoxide, and methane.

Future plans call for more laboratory and field work to study the utilization of these energy forms. The heat, naturally, would be used to produce steam for electrical generation. The fuel gases would be processed to air-condition buildings, drive cars, or whatever. But this processing would take place only after greatly increasing the volume of the gases by the addition of biomass—by dumping into the volcanic holes such things as old trees, logs, bark, sawdust, plant residue, garbage, trash—anything basically organic.

An even more daring plan to search out and capitalize on geothermal resources involves sending prospectors down to the rift areas in the Caribbean and the Pacific where the earth's crustal plates are pulling apart as much as 8 inches (20 centimers) a year. And if you will think back for a moment, you will recall that it's at the crustal plate boundaries that most of the world's hot spots occur. The prospectors sent to explore these rifts, however, would in no way resemble the backpacking old-timers who roamed the mountains and valleys hunting for geothermal clues with hand shovels and picks, bundles of dynamite, and mules. These would be crack teams of marine engineers, geologists, oceanographers, chemists, and physicists. These prospectors would sail off in computerized ships, dive down in specially designed submersibles, and hunt with underwater television color cameras.

Does this sound like a venture for the future? As a matter of fact, several such ventures are already on record. Since 1973, various ocean scientists have actually been exploring spreading sea-bottom rifts in tiny, deep-diving submersibles. The most enthusiastic promoter of these exploratory expeditions, which are funded by various governments, is Dr. Robert D. Ballard. A marine geologist at the Woods Hole Oceanographic Institution, Ballard has taken more dives to the bottom than any scientist alive, except perhaps Dr. John B. Corliss from Oregon State University. Never to be seen without his Cousteau-like cap, Ballard is excited by what may be going on in the rifts and is doing his best to find out.

In the photograph on page 110 you can see him inside the *Alvin* using the underwater phone to keep in touch with the mother ship *Lulu*. In that humming, clattering 23-foot (7-meter) submersible, Ballard, together with his pilot and the third member of the crew (neither shown here), will check and record instrument readings, snack on sandwiches and coffee, and otherwise busy himself during the two-hour journey to the bottom. At a depth of 12,000 feet (3,658 meters), the *Alvin* will perch on a convenient ledge in the Caribbean Cayman Trough, reach out a mechanical claw, and begin collecting rock samples from the jumbled terrain. And what a jumble the three men see through their viewports: cliffs and plains that are strewn with rocks and boulders, seamed with cracks under shifting sand and silt, and alive with bubbling hot springs. At various depths they will also glimpse sea creatures such as pink shrimp, giant oc-

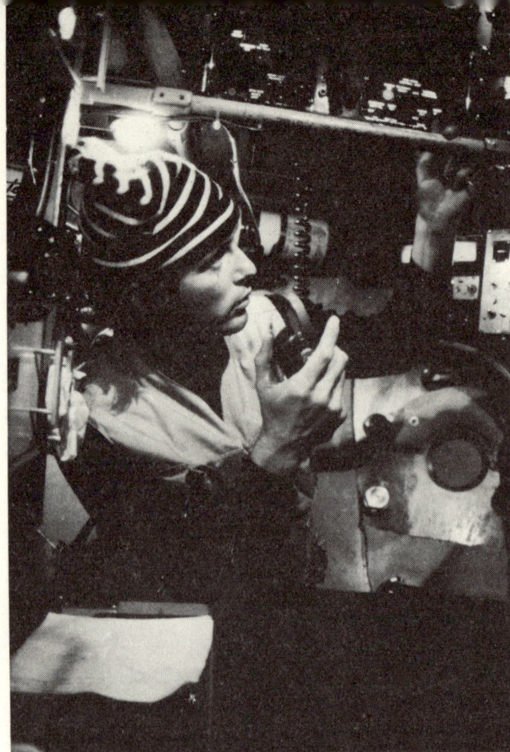

Left: Dr. Robert Ballard of the Woods Hole Oceanographic Institution in the personnnel sphere of the *Alvin,* using the underwater phone to communicate with the mother ship *Lulu.* (© National Geographic Society)

Right: The deep-sea submersible *Alvin.* (Woods Hole Oceanographic Institution, photo by John Porteous)

topods, rat-tail fish, corals, and more—all thriving down there in the dense darkness. Of course all is darkness below, but the *Alvin*'s blazing floodlights show the way.

Descriptions of the scenes some 2 miles (3 kilometers) below are not limited to the reports submitted by Ballard and the other scientists who've explored the rifts. What has really opened up this underwater geothermal story is the *Angus,* a 3-ton (2,721-kilogram) deep-sea sledge that's equipped with strobe lights, television color cameras, a bottom pinger (a sound-producing device), a temperature probe, and a temperature recorder.

The most spectacular observations ever were reported in 1979. At that time an expedition of American, French, and Mexican scientists, together with a filming crew from the National Geographic Society, explored the rift at the East Pacific Rise, 21 degrees north latitude, just southwest of Baja California. Among those heading this expedition were Dr. Ballard (as was to be expected) and marine biologist J. Frederick Grassle. Here, with the aid of computers, they positioned the mother ship *Melville,* a research vessel belonging to

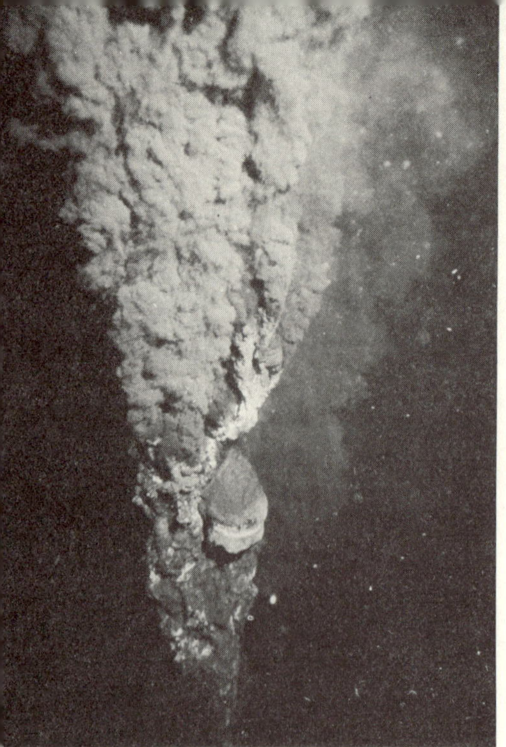
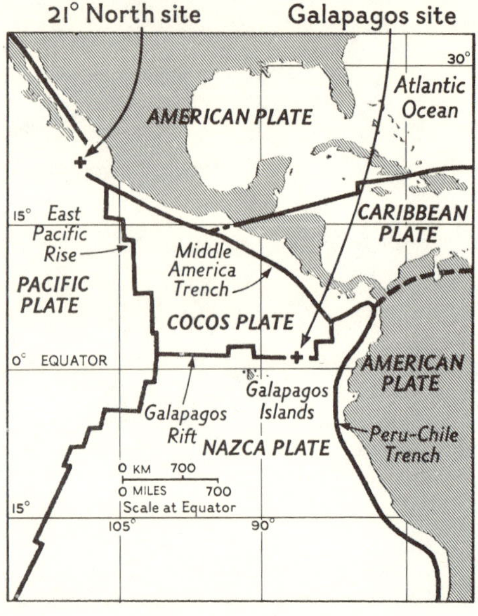

Left: A black smoker at the bottom of the sea at the East Pacific Rise. (Woods Hole Oceanographic Institution, photo by Robert Ballard) *Right*: At 21 degrees North latitude, the *Alvin* explores and the *Angus* films a portion of the spreading rift. (Adapted with permission from the National Geographic Society)

the Scripps Institution of Oceanography. Here they sent down the *Angus* to film a portion of the spreading rift and to take the temperature of the water. And here Ballard, invariably accompanied by two scientists, dived down in the *Alvin* to explore where no man had ever explored before. Their sea-bottom observations, together with the pictures taken by the *Angus,* recorded 1½ miles (2½ kilometers) below the surface of the calm blue Pacific, are now legendary.

They told about rows of "smokers," spewing dark clouds of fiercely hot water laden with minerals in solution. Smokers are geological chimneys formed when the hot geothermal fluids hit the cold sea water, which at that depth averages a near-freezing 35 degrees F

(2 degrees C). At that point, the dissolved minerals precipitate. They spatter the sides of the chimneys and fall in mounds of glittering crystals around them. Imagine seeing this process of ore formation—of zinc sulfides and iron and copper, as well as chalcopyrite (fool's gold), silver, lead, cadmium, and cobalt going on under your very eyes. It's like seeing Creation taking place.

They told about hot springs gushing out of the vents in the rift, some registering temperatures of 55 degrees F (13 degrees C), others of 660 to 750 degrees F (349 to 399 degrees C).

They told about the vents teeming with sea creatures: flowerlike sea anemones, foot-long clams and mussels and crabs, as well as red-tipped tubeworms that were 12 feet (3½ meters) tall and filled with bright red hemoglobin.

No one can minimize the scientific interest that has been generated by these findings, but you're probably wondering about their significance. A natural question is whether these hot fluids, bubbling out of the ocean bottom vent, could be important to our use.

According to Dr. Barnea, they could be. "Drilling in such areas," he says, "may provide significant information. It's probable that the

A cluster of red-tipped tube worms, some of them 12 feet (3½ meters) tall, found in the Galapagos Rift area. (Woods Hole Oceanographic Institution, photo by Dr. Kathleen Crane)

unmixed geothermal waters from the bottom of the sea may carry an abundance of valuable minerals and metals, as in the case of the Red Sea brines. And perhaps the geothermal fluids, before they are cooled by the sea water, may prove to be hot enough and pressurized enough to be of possible use in the future."

Of course, this is "tomorrow talk," but tomorrow will come whether anything constructive is done about these energy forms or not. Geothermal energy is so significant, so abundant, and so super-hot that its extraction and utilization are really something to look forward to. Instead of saying, "Pipe dream. It's not possible," let's tune in on Shakespeare's Hamlet, who reminded us that:

> There are more things in Heaven and Earth, Horatio,
> Than are dreamt of in your philosophy.

And so we come to the future. The future, of course depends on future technology that is efficient, durable, and inexpensive. In the meantime, prospectors and developers are moving ahead with the best of twentieth-century technology. Even so, they must start each venture with a search for geothermal deposits just as the old-time prospectors did. They search and ultimately they find. And as every one of them will tell you, with a twinkle in his eye:

> *I find steam in the ground*
> *And steam in the air*
> *And I look for hot water*
> *When steam is not there.*
> *If there's no hot water*
> *Then hot rock will do.*
> *It is there for the finding,*
> *It's all below you.*

Acknowledgments

There are, in all the world, only a few thousand geothermal experts, but they are a lively crew. They are hard-working, hard-talking men and women, ready to develop the planet's terrestrial heat and help solve the energy crisis. So many of them supplied me with papers and photographs and valuable advice that this book is actually a shared experience.

For their assistance to me in its preparation, my gratitude goes to the following people whom I interviewed in person in New York, Montreal, and Fullerton, California: Dr. Joseph Barnea, Senior Fellow, Project on the Future, United Nations Institute for Training and Research; Nicholas Lorimer, New Zealand Consulate; Dr. James McNitt, technical advisor, The United Nations; Dr. Tsvi H. Meidav, consultant in geothermal research and development; Steven Okulewicz, The College of Staten Island; Thomas Prendergast, The United Nations Library; Dr. Prem Saint, California State University; Gary Shulman, president, Geothermal Power Company Inc.

Gratitude also goes to the following for their ready assistance by way of telephone interviews: David N. Anderson, executive director, Geothermal Resources Council, California; Dr. Giancarlo Facca, registered geologist; Roger A. Fontes, California Energy Commisson; Dr. David B. Lombard, U.S. Department of Energy; Dr. Harold M. Stoller, Sandia National Laboratory; Dr. Stanley H. Ward, University of Utah Research Institute; Edward F. Wehlage, president, The International Society for Geothermal Engineering, Inc.

With no less appreciation I recognize the following for professionally helpful correspondence: Al Arenal and Lawrence P. Papay, Southern California Edison Company; Sylvia Briem, Iceland Tourist Bureau; Barbara Burress, Los Alamos Scientific Laboratory; Kathy Butler, U.S. Department of the Interior, Menlo Park, California; Pete Canizaro and Doug Drummond, Sperry Vickers; Gary Carlson and Thomas A. Ortiz, State of New Mexico Energy and Minerals Department; Vicky Cullen and Nancy Green, Woods Hole Oceanographic Institution; Dr. Ivan M. Dvorov, U.S.S.R. Academy of Sciences, Scientific Council for Geothermal Research; Bonnie Edens and Lindre Hogaboom, Rehabilitation Center, Hot Springs, Arkansas; Dr. Gian Carlo Ferrara, geochemist, Pisa, Italy; Dr. Milford R. Fletcher, National Park Service, New Mexico; The Honorable Sverrir H. Gunnlaugsson, Embassy of Iceland; Dr. I. C. Isdale, superintendent, Queen Elizabeth Hospital, Rotorua, New Zealand; João Silva Junior, San Miguel, Azores; Dr. Deane Kihara, University of Hawaii; Barry Lane, Union Oil Company, California; Dr. Paul J. Lienau, Oregon Institute of Technology; Dr. Baldur Lindal, Virkir Consulting Group Ltd., Iceland; Richard H. Maeder, superintendent, Hot Springs National Park, Arkansas; John J. McNamara, J. M. Energy Consultants, California; Michael Miller, former air specialist, and Paul W. Girard, Pacific Gas and Electric Company; Dr. Robert Morton, University of Texas; Nobuo Nagata, Japan Geothermal Energy Development Center, Tokyo; Jane S. Pratt and Jim Swanson, U.S. Geological Survey; Kathy Puckett, education program manager, Office of Energy, Boise, Idaho; David W. Reynolds, International Park Affairs Division, U.S. Department of the Interior; Diane Sakai, University of Hawaii; Jack S. Schneider, Office of Public Affairs, U.S. Department of Energy; Tetsumaro Senge, chairman, National Parks

   Acknowledgments

Association of Japan; Barbara Shattuck, National Geographic Society; Saretta Sparer and Carol Tscudi, Engineering Societies Library; Fred G. Stair, Colorado School of Mines; Kiyoshi Takahashi, superintendent of Nikko National Park Office, Nature Conservation Bureau of Environment Agency, Japan; Penny Toston, Geophysical Institute of Alaska; Harley T. Warren, chief, Engineering Services Branch, U.S. Department of the Interior; and Mary C. Woods, Division of Mines and Geology, State of California Resources Agency.

For a critical reading of the chapters that pertain to their special fields of expertise, my warmest thanks go to: Dr. Giancarlo Facca, Michael Miller, and Dr. Howard Wilcox.

And finally, a special thank you for reading the entire manuscript goes to: David N. Anderson, Dr. Joseph Barnea, Dr. Octave Levenspiel, Dr. David B. Lombard, and Dr. Tsvi H. Meidav, as well as to my husband, Oscar Goldin.

And I also want to gratefully acknowledge the permission to reproduce text, maps, diagrams, and photos supplied by the Director-General of the New Zealand Department of Scientific and Industrial Research (DSIR), by the Commissioner of Works and Development (MWD), New Zealand, and by the General Manager, New Zealand Tourist and Publicity Department (NPS).

Bibliography

Books

Armstead, H. C. H. *Geothermal Energy—Its Past, Present and Future.* Somerset, N.J.: Halstead Press, 1978.

Carr, Donald E. *Energy and the Earth Machine.* New York: Norton, 1976.

Clark, Wilson. *Energy for Survival: The Alternate to Extinction.* Garden City, N.Y.: Doubleday, 1975.

Geothermal Energy and Wind Power . . . Alternate Energy Sources for Alaska. Prepared by the Geophysical Institute, University of Alaska, and the Alaska Energy Office, William McConkey, director. Edited by Robert B. Forbes. April 1976.

Lienau, Paul J., and Lund, John W., eds. *Multipurpose Use of Geothermal Energy.* Proceedings of the International Conference on Geothermal Energy for Industrial, Agricultural and Commercial–Residential Uses, October 7–9, 1974. Klamath Falls, Oreg.: Geo-Heat Utilization Center, 1974.

Peele, R. *The Mining Engineers Handbook.* New York: Wiley, 1941.

Small, Hy Dee. *Nature's Teakettle, Geothermal Energy for the People.* West Covina, Calif.: Geothermal Information Services, 1973.

Wehlage, Edward F. *The Basics of Applied Geothermal Engineering.* West Covina, Calif.: Geothermal Information Services, 1976.

United States Government Publications

Abstracts: Hot Dry Rock Geothermal Energy. Los Alamos Scientific Laboratory, University of California at Los Alamos, N.M., under auspices of the United States Department of Energy, January 1978–August 1979.

Bedinger, M. S.; Pearson, F. J., Jr.; Reed, J. E.; Sniegocki, R. T.; and Stone, C. G. *The Waters of Hot Springs National Park, Arkansas—Their Origin, Nature, and Management.* U.S. Department of the Interior, Geological Survey, prepared in connection with the National Park Service, 1974.

Colp, John. L., and Stoller, Harold M. *The Utilization of Magma Energy— A Project Summary.* Presented at the Second Circum-Pacific Energy and Mineral Resources Conference, Honolulu, Hawaii, July 31–August 4, 1978. Albuquerque, N.M.: Sandia Laboratories, 1979.

Expanding the Hot Dry Rock Program. Reprinted from *The Atom,* Los Alamos Scientific Laboratory, June 1979. U.S. Government Printing Office, 1979.

"Geopressured Geothermal Resources: An Unconventional Energy Source." U.S. Department of Energy, 1979.

"Geothermal Energy in Hawaii." Honolulu: Hawaiian Natural Energy Institute, University of Hawaii, January 1978.

Hamilton, Warren. "Plate Tectonics and Man." *United States Geological Survey Annual Report.* U.S. Department of the Interior, 1976.

McNamara, John J., and Kaufman, E. L. "Hot Dry Rock Geothermal Resource Ownership and the Law." Paper LA–8027–HDR. Los Alamos Scientific Laboratory, University of California, September 1979. U.S. Government Printing Office, 1979.

Articles

Alexander, Tom. "The New Earth." *Smithsonian,* January–February 1975, pp. 1–22.

Anderson, Winslow, M.D. "A Visit to the California Geysers in 1888." Reprinted from *Mineral Information Service,* California Division of Mines and Geology, August 1969, with permission from the book *Mineral Springs and Health Resorts of California,* Bancroft Co., 1892.

Axtman, Robert C. "Environmental Impact of a Geothermal Power Plant." *Science,* Vol. 187, No. 4179 (March 7, 1975), pp. 795–803.

Ballard, Robert D. "Window on the Earth's Interior." *National Geographic,* Vol. 150, No. 2 (August 1976), pp. 228–249.

Ballard, Robert D., and Grassle, Frederick J. "Return to the Oases of the Deep." *National Geographic,* Vol. 156, No. 5 (November 1979), pp. 689–705.

Barnea, Joseph. "Geothermal Power." *Scientific American,* Vol. 226, No. 1 (January 1972), pp. 70–77.

Birsic, Rudolph J. "What's New at The Geysers, U.S.A." *Geothermal World Directory,* 1978/79, pp. 289–292.

Boldizsar, Tibor. "Terrestrial Heat and Geothermal Energy Production." *Geothermal World Directory,* 1978/79, pp. 360–391.

Britton, Peter. "Geothermal Goes East." *Popular Science,* February 1979, pp. 66–69.

Cloud, Wallace. "The Ship That Digs Holes in the Sea." *Popular Mechanics,* March 1969, pp. 108, 111, 236.

Colp, John L., and Brandvold, Glen E. "The Sandia Magma Energy Research Project." *Geothermal World Directory,* 1978/79, pp. 424–431. (Advanced Energy Projects, Sandia Laboratories, Albuquerque, N.M.)

Dvorov, Ivan M., and Ledentsova, Nora A. "Utilization of Geothermal Water for Domestic Heating and Hot Water Supply." Scientific Council on Geothermal Research, Academy of Sciences of the U.S.S.R., All-Union Scientific Research Institute of Complex Fuel and Energy Problems, 1975, pp. 2109–2116.

Ellis, A. J. "Geothermal Systems and Power Development." *American Scientist,* Vol. 63, No. 5 (September/October 1975), pp. 510–521.

Facca, Giancarlo. "The Structure and Behavior of Geothermal Fields." *Unesco, Geothermal Energy* (Earth Sciences 12), 1973, pp. 61–69.

"The First Geopressure Well." *Important for the Future,* Vol. 2, No. 2 (April 1977), pp. 6–7.

Forbes, Robert B., and Biggar, Norma. "Alaska's Geothermal Resource Potential." *The Northern Engineer,* Vol. 5, No. 1 (1973), pp. 6–10.

George, Uwe. "An Ocean Is Born." *Geo-Collectors Edition,* 1979, pp. 93–112.

Henahen, John F. "Geothermal Energy . . . The Prospects Get Hotter." *Popular Science,* November 1974, pp. 96–99, 142–143.

Holden, William M. "Underground Steam." *Popular Science,* November 1968, pp. 83–86.

Holland, M. B. "Power from the Earth." *The Chartered Mechanical Engineers Journal* (London), November 1978, pp. 40–46.

Holt, Ben. "Geopressured Resource: A Sleeping Giant." *Geothermal Energy,* Vol. 5, No. 11 (November 1977), pp. 30–32.

Kerr, Richard A. "Glomar Explorer: New Era in Deep Sea Drilling?" *Science*, Vol. 200 (June 16, 1978), pp. 1254–1255.

Lautman, Maurice F., M.D. "Hydrotherapy in Arthritis." *Archives of Physical Therapy, X-Ray, Radium*, Vol. XV (February 1934), pp. 107–110.

Leonard, Lee. "What's Old in Geothermal Energy?" *The Northern Engineer*, Vol. 6, No. 4 (1974–75), pp. 3–7.

Lindal, Baldur. "Magma Heat for House Heating." *Important for the Future*, Vol. 2, No. 2 (April 1977), p. 6.

Lund, John. "Direct Utilization—The International Scene." *Geothermal World Directory*, 1978/79, pp. 311–323.

Meidav, Tsvi. "Geothermal Opportunities Bear a Closer Look." *Oil and Gas Journal*, May 13, 1974, pp. 102–106.

Meidav, Tsvi, and Meidav, Mae. "The Direct Heat Utilization of Geothermal Energy on the Island of San Miguel." Geothermal Resources Council *Transactions*, Vol. 3 (September 1979), pp. 31–33.

Miller, Charles A. "Geothermal Power Is Here!" *Mechanix Illustrated*, Vol. 75, No. 613 (June 1979), pp. 31–33.

O'Keefe, William. "Geothermal Power: Sleeping Giant Stirs But Will Require Years to Waken Fully." *Power*, April 1973, pp. 32–36.

"P G and E Electric Supply, Planning, Process, and Current Supply Plan." Pacific Gas and Electric Company, October 1979 (Volume 11, appendices), pp. A1–A15.

Rea, Kenneth H. "Environmental Investigations Associated with LASL Hot Dry Rock Geothermal Energy Development Project." *Geothermal World Directory*, 1978/79, pp. 401–423.

Rowley, John C. "Geothermal Energy Development." *Physics Today*, Vol, 30, No. 1 (January 1977), pp. 36–46.

Saint, Prem K. "Volcanoes and Geothermal Energy in New Zealand," *Geothermal Energy*, Vol. 5, No. 3 (March 1977), pp. 8–17.

Schlec, Susan. "21° North." *Science 80*, November/December 1979, pp. 37–46.

Schuster, Eric J. "The Search for Hot Rocks: Geothermal Exploration, Northwest." *Pacific Search*, May 1973, pp. 1–3.

Scully, Francis J., M.D. "The Role of Radio Activity of Natural Spring Waters as a Therapeutic Agent." *Journal of the Arkansas Medical Society*, Vol. XXX (March 1934), pp. 206–214.

———. "Spa Therapy of Arthritis and Rheumatic Disorders." *Archives of Physical Medicine*, Vol. XXVI (April 1945), pp. 233–238.

Tester, J. W., and Smith, M. C. "Energy Extraction Characteristics of Hot

Dry Rock Geothermal Systems." *Geothermal World Directory,* 1978/79, pp. 340–347.

Weaver, Kenneth F. "The Power of Letting Off Steam." *National Geographic,* Vol. 152, No. 4 (October 1977), pp. 566–579.

West, Susan. "DSDP: 10 Years After." *Science News,* Vol. 113, No. 25 (June 24, 1978), pp. 408–410. (DSDP means Deep Sea Drilling Project.)

————. "Smokers, Red Worms and Deep Sea Plumbing." *Science News,* Vol. 117, No. 2 (January 12, 1980), pp. 28–30.

Papers

Anderson, David N. "The Most Promising Geothermal Fields in the Western United States (Including the Geysers Geothermal Field)." Presented at the Geothermal Resources Council Special Short Course No. 8, South San Francisco, May 8–9, 1979.

Barnea, Joseph. "Economics of Multi-Purpose Uses of Geothermal Resources." In *Proceedings of the International Conference on Geothermal Energy for Industrial, Agricultural, Commercial—Residential Uses,* Klamath Falls, Oregon, October 1974, pp. 209–216.

————. "Geothermal Minerals—The Neglected Minerals." In *UNITAR Small Scale Mining of the World Conference,* Jurica, Mexico, November 1978, pp. 38–56.

Bolton, R. S., and Studt, F. E. "Investigation and Development of New Zealand's Geothermal Resources." Ministry of Works and Development and the Department of Scientific and Industrial Research, 1977.

"Current Trends of Geothermal Development and Utilization in Japan." Japan Geothermal Energy Center, August 1979.

"Geothermal Energy in New Zealand." Ministry of Works and Development, April 1974.

Johnson, William C. "Aquaculture Using Geothermal Energy." Oregon Institute of Technology, 1978.

Lienau, Paul J., and Lund, John W. "Utilization and Economics of Geothermal Space Heating in Klamath Falls, Oregon." Geo-Heat Utilization Center, Oregon Institute of Technology, 1977.

Lund, John W. "Direct Utilization—The International Scene." Geo-Heat Utilization Center, Oregon Institute of Technology, 1977.

McMillan, D. A. "Economics of The Geysers Geothermal Field, California." United Nations Symposium on the Development and Utilization of Geothermal Resources, Pisa, Italy, 1970.

Meidav, Tsvi. "Overview of Worldwide Geothermal Developments." Geothermal Resources Council Workshop, San Diego, California, November 1978.

"The Mineral Waters of São Miguel." Comissão Regional de Turismo Das Ilhas de São Miguel e de Santa Maria, August 1979.

Shulman, Gary. "The Geothermal Power Monoblok for Small Scale Mining." In *UNITAR Small Scale Mining of the World Conference,* Jurica, Mexico, November 1978, pp. 478–488.

Shupe, John W., and Yuen, Paul C. "Geothermal Energy in Hawaii—Present and Future." Circum-Pacific Energy and Mineral Resources Conference, Honolulu, Hawaii, August 1978.

UNITAR Conference on Long-Term Energy Resources, Montreal, Canada, November/December 1979. The following papers given there were helpful in the preparation of this book:

Barnea, Joseph. "Long-Term Energy Resources and the Energy Outlook."

Boldizsar, Tibor. "Geothermal Energy for Space Heating."

Cowhey, Peter, and LeGates, Richard. "The Use of Geothermal Energy to Develop Islands."

Espinosa, Hector Alonso. "Actual Situation and Future Programme of the Geothermal Potential of the Mexican Republic."

Facca, Giancarlo. "Geothermal Energy Development: A Historic Summary."

Ferrara, Gian Carlo. "Geothermal Energy for Electrical Production: Present Status and Future Prospects in Italy."

Hendron, Robert H. "Energy from Hot Dry Rock."

Jaffé, F.; Rybach, L.; and Vuataz, F. D. "Exploration of Low Enthalpy Geothermal Energy in Switzerland."

Lindal, Baldur. "Geothermal Energy from Warm Water."

McNitt, J. R. "Current Activities of the United Nations in the Development of Geothermal Resources."

Meidav, Tsvi. "Exploration of Low and Intermediate Enthalpy Geothermal Resources."

———. "Trends in Geothermal Exploration Technology."

Yora, Mitsuo. "Development and Utilization of Geothermal Energy in the Hot Water Fields of Japan." Geothermal and Energy Development Center, 1978.

Ward, Stanley H. "Technology of Utilizing Geothermal Energy." Presented to the Conference on Health Implications of New Energy Techniques, Park City, Utah, April 1979.

Waring, Gerald A. "Thermal Springs of the United States and Other Countries Around the World—A Summary." Geological Survey Professional Paper 492. Washington, D.C.: U.S. Government Printing Office, 1965.

Brochures, Bulletins, and Booklets

"Annual Information on Development and Utilization for Geothermal Energy in Japan." Tokyo: Japan Geothermal Energy Association, 1978.

Bedinger, M. S. *Valley of the Vapors: Hot Springs National Park.* Philadelphia: Eastern National Park and Monument Association, 1974.

"Geothermal Electric Power Plants of Larderello and Monte Amiata: Electric Power from Engogenous Steam." Seri Grandi Impianti ENEL. Pisa, Italy: Research Central Management, Public Relations and Press Office, 1976.

Geo-Heat Utilization Center Quarterly Bulletin. Oregon Institute of Technology. May 1975, January 1976, November 1977, June 1979.

Hot Springs Rehabilitation Center. Arkansas Rehabilitation Service, Department of Human Services, 1977.

Magnusson, Sigurdur A. "Iceland—Country and People." Reykjavik: Iceland Review, 1978.

"New Zealand Geothermal Development." Information and Publicity Group of the Ministry of Works and Development, April 1974.

Power from the Earth: The Story of the Wairakei Project. Wellington, N.Z.: A. R. Shearer, Government Printer, 1970.

"The Present Status of Geothermal Development and Utilization in Japan." Japan Geothermal Energy Development Center, August 1978.

"The Present Status of World Geothermal Development." Energy and Mineral Development Branch, United Nations, May 1975.

Reed, A. W. *Power from the Earth at Wairakei.* Wellington, N.Z.: A. H. and A. W. Reed Ltd. and Longman Paul Ltd.

Index

First Geopressured Geothermal
Conference, 102
Flash system, electricity produced
with, 71–73, 75
Forbes, Robert B., 88
Friction, as source of heat, 17
Fumaroles, 21, 40, 50, 61

Geopressured zones, 100–103, 106
Geothermal energy
defined, 12
low cost of, 13–14, 49–50, 94
problems in harnessing the use
of, 39, 45–47, 49, 62–67, 70
sequential (cascading) use of,
94–95
use of for electricity, 13, 37–40,
44–51, 52, 55, 61–70, 71–77
uses of, nonelectric (direct), 13,
91–95, 98–99
See also Geysers, The; Iceland;
Italy; New Zealand
*Geothermal Energy and Wind
Power*, 88
Geothermal energy park (Azores),
95–99
Geothermal hot water, high-temperature, 52–55
drilling for, 60–61, 62–63
techniques for locating, 55–56,
58–60
See also Steam, wet
Geothermal hot water, low-temperature, 71, 78
nonelectrical uses of, 91–93
and the production of electricity,
71–77
Geothermal underwater exploration, 108–114
Geysers, 12, 17, 21, 22, 40
Geysers, The (steam fields in Cal.),
40–50, 55, 65, 94
Grassle, J. Frederick, 110
Gravity studies, as an exploratory
technique for locating geothermal fluids, 58

Greeks, ancient, 25
Greenhouses, geothermally heated,
83, 93, 95, 99

Hawaii Institute of Geophysics, 77
Heat exchanger
downhole, 89–91
surface, 88–89
Heber (Cal.), power plant in, 75
Heimaey (Iceland), volcanic heat
utilized in, 106
Hot dry rocks, 100, 103–106
Hot spots, 17, 21
Hot springs, 12, 17, 21, 22, 23, 29,
50, 81, 87
bathing in, 25, 27
chemical composition of, 30–31
cooking with, 23, 25
heating homes with, 23, 25
and Indians, American, 24, 40
therapeutic value of, 24–25, 27,
30–33
See also Spas
Hot Springs (Ark.), 24, 32
Hot water, geothermal. *See* Geothermal hot water, high-temperature; Geothermal hot
water, low-temperature
Hveragerdi (Iceland), 83
Hydrofracturing techniques, 104
Hydrogen sulfide problem, 45–47,
49
Hydros, 31
See also Hot springs; Spas
Hydrotherapy, 32–33

Iceland, geothermal energy used
in, 14, 80–87, 106
Idaho, geothermal energy used in,
75, 95
Illness, therapeutic value of hot
springs as treatment for, 30–31
India, huts heated by hot springs
in, 31
Indians, American, and hot springs,
24, 40

Indonesia, electricity generated
    from geothermal hot water in,
    36, 52, 71
International Conference on Long-
    Term Energy Resources
    (1979), 91
Isdale, I. C., quoted, 32–33
Italy, geothermal energy used in,
    14, 37–40, 55, 61

Japan
    aquaculture in, 93
    geothermal energy used in, 14,
        36, 40, 50–51, 93
    hot-spring bathing in, 27, 29–31
Jigoku, 31
Jones, Paul, 102

Kamiyu, 27
Karshner, J. F., 33
Kilowatt hour (KWH), 49, 50
Klamath Falls (Oreg.), downhole
    heat exchangers used in,
    89–91, 94

Larderello. *See* Italy
Lindal, Baldur, 91
Los Alamos Scientific Laboratory
    (LASL), 103–105
Lulu (ship), 109

Magma, 12, 15, 53
    tapping energy from, 106, 108
Magma Power Company, 45, 75
Magmamax, 77
Magnetics, as exploratory tech-
    nique for locating geothermal
    fluids, 58–59
Manley, Frank, 33
Manley Hot Spring Plantation Re-
    sort, 33–34
Mantle of the earth, 14, 15
Megawatt, defined, 39–40
Meidav, Mae, 95, 98–99
Meidav, Tsvi, 95, 98–99
Melville (ship), 110

Mexico, 31
    geothermal energy used in, 14,
        52, 70, 95
Miller, Michael J., quoted, 47, 49
Moho, 16
Mohorovičić, Andrija, 16

National Geographic Society, 110
New Mexico, geothermal energy
    used in, 77, 103, 106
New Zealand
    geothermal energy used in, 14,
        52, 61–70, 94, 95
    hot springs used by Maoris in, 25
    Queen Elizabeth Hospital in,
        32–33
Nikko National Park (Japan), 29–30

Pacific Gas and Electric Company,
    45, 47, 49
Pacific Ring of Fire, 21, 50
Paratunka (Siberia), binary pilot
    project in, 75
Philippines, electricity generated
    from geothermal energy in, 36,
    71
Power from the Earth at Wairakei
    (Reed), 70
Puna (Hawaii), hot-water deposit
    at, 77

Queen Elizabeth Hospital (Rotorua,
    N.Z.), 32–33

Radioactive elements in rocks, 16
    decay of, 16–17
Reed, A. W., quoted, 70
Refrigeration powered by geother-
    mal energy, 91, 94, 98, 99
Reservoirs, geothermal. *See* Geo-
    thermal hot water, high-tem-
    perature; Geothermal hot
    water, low-temperature; Steam
Reykjavik (Iceland), geothermal hot
    water used in, 83
Ring of Fire, Pacific, 21, 50
Rock layers of the earth, 52–54, 58